TRANSFER OF AGRICULTURAL TECHNOLOGY; STRUCTURAL AND FUNCTIONAL LINKAGES

TRANSFER OF AGRICULTURAL TECHNOLOGY; STRUCTURAL AND FUNCTIONAL LINKAGES

A Study of Improved Rice Varieties

O. ABDUL RAHIMAN KUNJU

Foreword by
C. PRASAD
Deputy Director General (AE), ICAR
NEW DELHI

CONCEPT PUBLISHING COMPANY, NEW DELHI 110059

The Indian Council of Social Science Research gave the financial support towards the publication of this work. However, the responsibility for the contents of the book — facts, opinions, conclusions, etc. — is entirely that of the author and not of the ICSSR.

ICSSR Consultant: **Dr. Daulat Singh**

ISBN 81-7022-394-6

First Published 1992

Published and Printed by
Ashok Kumar Mittal
Concept Publishing Company
A/15-16, Commercial Block, Mohan Garden
NEW DELHI 110059 (India)

Lasertypeset by
Microtech Advance Printing Systems (Pvt.) Ltd.
H-13, Bali Nagar,
NEW DELHI 110015

To

MY ELDEST SON

LATE A. MOHAMMED RAFFI

FOREWORD

AGRICULTURE WILL continue to be the main occupation of majority of the rural population for a long time to come. While we have brought Green Revolution in the country, the growing population demands a lot more efforts in the field of research, education and extension. In view of large number of productive technologies available everywhere in the country, it is high time that Transfer of Technology gets priority over research or education. Unless massive efforts are made and proportionate investments are provided by the governments, the task of producing 225 million tonnes of foodgrains by A.D. 2000 may not be realised.

There are five extension systems/sub-systems namely: (i) First-line Extension Education System; (ii) Main extension system of the Ministry of Agriculture/State Department of Agriculture; (iii) Special thrust programmes of the Crop Division of the Ministry of Agriculture; (iv) Technology Mission on Oilseed Production; and (v) Department of Rural Development of the Ministry. The first-line extension education role is basically the institutional role in Transfer of Technology. This role is extremely important as catalytic force so that the time-lag between technology generation and technology adoption is reduced. Lately, the concept of First-line Extension Education of the ICAR including State Agricultural Universities and the ICAR Research Institutes have greatly changed and the focus is on the following specific objectives:

1. Promptly demonstrating the latest agricultural technologies to the farmers as well as the extension workers of the State Departments of Agriculture/Non-government Organization with a view to reduce the time-lag between the technology generation and its adoption.
2. Testing and verifying the technologies in the socio-economic conditions of the farmers and identifying the constraints.
3. Getting the first-hand scientific feedback by the scientists, in order to keep abreast with the performance of the technologies

and the farming problems, and reorient their research, education and training programmes accordingly.

4. Providing training and communication supports to the State Departments of Agriculture/NGOs.
5. Based on field-work and experience, develop extension models to be adopted by general extension systems for large-scale multiplication.
6. Promoting research in extension and studies including comparative studies of extension systems in different parts of the world.

It is essential, therefore, that the extension scientists including subject-matter specialists pay closer attention on these objectives and insist on quality extension work so that their work is much different quality-wise than the general extension system of the State Department of Agriculture. The work of the first-line extension system should be treated basically as an 'action research' which should provide guidance and models to the general extension system. The farming system approach in research and extension must be promoted especially in view of large number of farmers being small and marginal.

Rice is a number one crop in the country and more so in Kerala State. The research work on structural and functional linkages in transfer of technology of improved rice varieties released by the Kerala Agricultural University – a systems analysis has been a commendable work. I am glad that this is being published in the form of a book so that the work could find a rightful place in the publication world.

I am sure this publication will be useful to researchers as well as extension workers alike.

Deputy Director General (AE) C. PRASAD
Indian Council of Agricultural Research
Krishi Anusandhan Bhavan, Pusa
NEW DELHI-110012

ACKNOWLEDGEMENTS

IT GIVES me immense pleasure to place on record my indebtedness to Dr. A.M. Tampi, Coordinator, Advanced Centre for Training in Plantation Crops for his sincere help and constant encouragement which enabled me to complete this research work.

I express by gratitude to Dr. A.G.G. Menon, Director of Extension, KAU, Dr. G.T. Nair, Professor of Extension, Prof. P.V. Prabhakaran, Professor and Head Department of Agricultural Statistics and Dr. George Koshy Professor of Entomology, College of Agriculture, Vellayani for their valuable suggestions during the course of this research study.

I am deeply indebted to Dr. C. Bhaskaran, Associate Professor of Extension, College of Agriculture, Vellayani for the sincere help rendered to me at all stages of this research. His trenchant and penetrating comments helped me a lot to improve upon this publication.

I am grateful to Shri C.E. Ajith Kumar, Junior programmer of the Department of Agricultural Statistics, College of Agriculture, Vellayani for the help in computer programming and analysis of data and Sri M.M. Rasi for the graphic assistance.

My thanks are also due to all my colleagues at the college of Agriculture, Vellayani, who rendered help in the completion of this research work.

I express by gratitude to Dr. C. Prasad, Deputy Director General, Indian Council of Agricultural Research, New Delhi for blessing this publication by way of writing a foreword.

I am also obliged to the Kerala Agricultural University for giving me permission to bring out this publication.

O. Abdul Rahiman Kunju

CONTENTS

LIST OF TABLES

LIST OF FIGURES

1

INTRODUCTION

THE PROCESS of technology transfer in agriculture is a long and intricate one, involving various institutions and different stages from research stations to the farmers' fields. The mechanics of transferring the technologies to the actual users is now gaining renewed attention.

The process of transfer of technology (TOT) cannot be examined in isolation. The effectiveness of the TOT is greatly influenced by the linkage between and among those responsible for technology generation (research sub-system) and technology transfer (extension sub-system), those who actually put to use the technology (client sub-system) and those who act as catalysts in the adoption of modern technology by supplying the required inputs (input sub-system). Thus TOT of any improved practice calls for a concerted and coordinated effort and mutual understanding between and among the above four sub-systems involved.

In this context, it was felt that systems analysis would provide the framework for consideration of all the component sub-systems involved in the TOT process. The process of TOT could be examined wholly taking the process *per se* as the major system and the research, extension, client and input sub-systems as its components.

The need for strengthening the linkages between and among the different sub-systems has always been felt, more intensely after the initiation of the ambitious High Yielding Varieties (HYV) Programme during the 1960s and the consequent green revolution. As a matter of fact, the ignition for the green revolution was provided by the research sub-system. This brought the State Agricultural Universities (SAUs) and the Research Stations of Indian Council of Agricultural Research (ICAR), Council of Scientific and Industrial Research (CSIR), Commodity Boards etc. to a greater limelight and this led to a scramble among the clientele for new technology, inputs and services. As a corollary to this, a host of organisations engaged in the TOT began to gear up their efforts and thereby, accelerated the process of green revolution. Simultaneously, there was a proliferation of agro-service centres and agencies for distribution of seeds, fertilizers and pesticides. The mass media devoted more time and space after the advent of the green revolution than during the pre-green revolution phase.

With the establishment of Agricultural Universities, research in totality and technical backstop of extension partially were transferred to the University. It is worth pointing out that the duty of the research system does not end with the release of improved or "miracle" seeds and/or appropriate technologies. Any technological innovation brings with it, a host of new problems which necessitate further research, both biological and social. For understanding these problems and for offering solutions to them, research will have to work in unison with extension. The Training and Visits (T&V) system started under the National Agricultural Extension Project (NAEP) forged these linkages at various levels. The National Agricultural Research Project (NARP) started in the 1980s, which also is a major effort by the ICAR in this direction, strengthened the regional research capabilities of SAUs. Thus the two projects – the NAEP and NARP – together paved the way for systematic and formalised linkages and coordination between and among the various sub-systems.

At the client system level, the expanding and dynamic information matrix required for modernising the agricultural practices of the farmer need to be channellised by sources lying much beyond the frontiers of his home. The farmer needs the appropriate technology and the recommended input for increasing production. Information about the economic, organisational and socio-cultural factors which influence the nature and efficiency of his farm enterprise should also be available to him. This again calls for an efficient functional linkage between the farmer sub-system and the other sub-systems. Under the new extension system, the process of TOT to the farmers' fields takes place through regular time-bound visits by the village level Agricultural Demonstrators (ADs) who are trained by the Subject Matter Specialists (SMSs) every fortnight. The visits of the ADs are supervised by the Agricultural Officers (AOs) working in the Agricultural Extension (AE) Units and also the SMS. The SMSs in turn, have interaction with the scientists every month at the monthly workshops. Thus structured and systematic functional linkages between and among the research extension-client sub-systems have been ensured. Simultaneously, there is regular feedback of field problems to the research sub-system enabling it to make research more problem-oriented and location-specific. The farm trials laid out in the farmers' fields and the field visits made jointly by the research and extension sub-systems again bring the extension, research and client sub-systems close together.

The missing linkage between the extension and input sub-systems is being welded through organising national input fortnights. Emphasis is given to organising supply of various production inputs like seed, fertilizer, credit etc. during the input fortnight celebrations. In the context of the new agricultural strategy, the role of the institutions which provide the necessary material and cash inputs cannot be over-emphasised. The dealers' trainings organised by the Sub-Divisional Agricultural Officers (SDAOs) and the involvement of the input agencies in seminars, field days etc. organised by research and extension sub-systems link the input sub-system to the other three sub-systems.

The Problem

Rice is the major food crop of Kerala, covering 23 per cent of the total cropped area in the State (Farm Guide, 1988)*. While the State is reeling under acute shortage of rice to meet the internal demand, the production of the rice crop remains stagnant causing concern to the people of Kerala. This problem has to be solved urgently in the larger interest of the States' agricultural economy. The Kerala Agricultural University (KAU) since its inception in 1972 has initiated research projects to boost up rice yields in the State. In this pursuit, the KAU has so far released 14 rice varieties. However, available statistics show that only about 25 per cent of the rice area in the State is being cultivated with the HYVs of rice (Farm Guide, 1988) in spite of the presence of reputed Agricultural Research Stations, a reorganised Department of Agriculture, an enlightened client system and a network of input agencies in the State. This shortfall in coverage could partly be due to the weak structural and/or functional linkages that exist between and among the research, extension, client and input sub-systems. Since comprehensive and systematic analysis to study the strength and weakness in the linkage between and among these sub-systems has not been done so far, the present investigation was designed with the following specific objectives.

---

* Farm Information Bureau, Government of Kerala.

(1) To study the extent of linkage activities performed by, between and among the research, extension, client and input sub-systems in the TOT of the improved rice varieties released by the KAU.
(2) To analyse the factors associated with the linkages between and among the different sub-systems in the TOT of improved rice varieties released by the KAU.
(3) To identify the constraints in evolving, spreading and adopting improved rice varieties as perceived by the research, extension, client and input sub-systems.
(4) To suggest a model of linkages between and among the research, extension, client and input sub-systems in the TOT in Agriculture.

Scope of the Study

The study attempts to explore the extent of linkage activities performed by the research, extension client and input sub-systems with particular reference to the TOT of improved rice varieties released by the KAU. A study on the linkages between and among the sub-systems involved in the TOT process would bring to focus the lacunae in the TOT process and identify the weak links. Based on the results, appropriate measures could be suggested to overcome the limitations and strengthen the weak links.

2

THEORETICAL ORIENTATION

A REVIEW of previous research studies helps in delineating new problem areas and provides a basis for developing a theoretical framework for the present study. This will also help in operationalising the variables and concepts on the basis of which required data could be collected. In accordance with the specific objectives set, the review of literature related to the study is furnished under the following sub-heads:

2.1 Concept of linkage
2.2 Concept of structural and functional linkages
2.3 Importance of structural and functional linkages
2.4 Concept of transfer of technology
2.5 Concept of systems analysis
2.6 Studies on the linkages for transfer of technology
2.7 Constraints in the transfer of technology process
2.8 Conceptual model of the study

2.1 Concept of linkage

According to *Webster's Dictionary* the term 'linkage' means a series or systems of links. Literally, link means anything serving to connect or tie.

According to Sharma (1982) the word linkage also means to chain in one way or other and a chain may have two functions, namely,

i. to keep a check, to control or to limit movement up to a certain point,
ii. to facilitate co-ordinated/controlled movement for advancement or for fulfilment of some desired or productive purpose.

The above two roles may be expected of a linkage too. However, the second role is more applicable in the context of the transfer of technology.

In the context of communication, the term link is used to indicate the existence of communication facilities between two points.*

Binding together or uniting are synonymous to the words connecting or tying. These are two more expressions that could be added in conceptualising a linking.

Havelock *et al.* (1969) stated that linking roles had been closely related to the dissemination and utilisation of knowledge. They were of the opinion that the natural starting point for the discussion of linking role is knowledge gap, the situation for which linkage is required. This gap could be bridged effectively if additional persons or groups are interposed like the researchers and clients, these additional intermediaries being the specialists in the process of linking itself.

The authors listed nine types of linkage namely, (a) conveyor, (b) consultant, (c) trainer, (d) leader, (e) innovator, (f) defender, (g) knowledge builder, (h) practitioner and (i) user. A brief description of each is given below:

a) Conveyor

It is the most rudimentary and simple linker concept. The conveyor takes knowledge from experts and passes it on to non-expert potential users. The knowledge of source could be in the form of research data, information derived from research, packaged knowledge derived generally from scientific knowledge in the form of circulars, printed materials and training programmes.

b) Consultant

Consultant is a facilitator, helper, objective observer and specialist in how to diagnose needs, identify resource, and retrieve knowledge from expert sources. He tells us 'how' in contrast to the conveyor who tells us 'what'.

---

* *McGraw-Hill Encyclopaedia of Science and Technology*, Vol. 7, 1977.

c) Trainer

The trainer works on the assumption that underlies much of formal education, viz., a body of knowledge received and stored for future use. The trainer is an expert who is capable of transferring large quantities of knowledge and/or complex skills, but he does not typically convey this knowledge to people who are in work setting.

d) Leader

There are great number of roles which create effective linkage through power or influence within the 'receivers' own group. These various role types are discussed under the designation 'leaders'. There is enough evidence that formally constituted functionaries administrators, directors, supervisors and presidents do have major influence on utilisation of new ideas. The leader linking role also includes the concept of opinion leaders.

e) Innovator

The innovator is a linker in three ways: (i) he may be a latent opinion leader perhaps through success and prosperity which may result from being an innovator; (ii) the innovator serves as linker being demonstrator and quasi-opinion leader for the real opinion; and (iii) the innovator may become an effective linker through the active advocacy of innovation.

f) Defender

The linkage role played by defender in the utilisation of knowledge is the communication of negative information. All new ideas and changes bring with them some problems and some reasons and that is why all adoption is not possible. Some client systems are too open to change and adoption of new ideas, too unaware of the pitfalls of innovation and too much vulnerable to danger. The defender is always on watch for these dangers and ready to sound the trumpet to awaken the public.

g) Knowledge Builder

The linkage role of knowledge builder includes the part played by scientists, scholars, engineers and other knowledge builders in the process of dissemination and utilisation. The linking roles performed by the knowledge builders however, depend on, how they are positioned in the social system and how they are used by others.

h) Practitioners

They are also linkers of knowledge to the ultimate consumers by providing specialised sources. Over emphasis on professionalism and specialisation by practitioners may also weaken the linkage to the consumer.

i) User

Users acting themselves as linkers is rather uncommon. In a few cases one may find that the most sophisticated farmer with long experience, extension exposure and training does not rely on the officers of the department of agriculture. Even in a very advanced system, this pattern is an exception and not a rule.

Axinn and Thorat (1972) described linkage as clusters of channels which connect one major component with other major components.

Sharma (1982) stated that linkage has an element of co-ordination. Co-ordination is a process meant for accomplishing the desired interaction, co-operation and/or collaboration. All these involve some element of integration too and working in unison with one focus in view or for achieving some common target, goal or objectives.

From the above discussion, a simpler expression for linkage could be derived: A working relationship or functional relationship between and among sub-systems, for a purposive interaction. This relationship may be very simple, intricate/or inextricable. When the relationship has to be of a complex nature, it has to be in the form of a series or system of link.

2.2 Concept of Structural and Functional Linkages

2.2.1 *Structural Analysis*

The word 'structure' literally means arrangement of parts of anything. According to Mann (1983), structure is a concept widely used in Sociology, but in senses either too specific or too general to permit simple agreed definition. Primarily, structure is distinguished from simple aggregation or conglomeration of elements by features loosely grouped into three linked categories, viz., (i) a set of elements is connected by relations to form a whole, with new properties not possessed by the elements alone, (ii) substitution of new elements for old is possible without necessarily altering the nature of the structure; and (iii) the ways in which the structure may be transformed serve to define it in a way equivalent to its internal relations.

The new properties associated with a structure are seen as dependent more on the nature of the relations linking it together than on the substantive nature of the parts. Substitution, as long as their relations are preserved, will have little effect on the structural properties.

From the structural point of view, a part is specified by its position within the whole. Transformation of one or more areas of the internal relations will be systematically linked to transformation of remaining areas. The nature of these links can give an exact image of the structure.

Applying this connotation for the purpose of the present study, transfer of technology of improved rice varieties forms the whole. The change agent provides communication link between a resource sub-system of some kind and a client sub-system. One of the main roles of a change agent is to facilitate the flow of innovation from the resource sub-system to the client sub-system. This type of communication to be effective, the innovation must match the needs of clients. Also the required inputs must be made available. Thus the different sub-systems are structurally linked, at the same time keeping their identity, together to form the whole viz. the system of transfer of technology.

2.2.2 *Functional Analysis*

According to Singh (1980), it was Leibniz who first introduced the word function in the mathematical context, where, in a more extended sense, it meant "functional inter-dependence" and "functional relations".

Merton (1957) viewed that functional analysis in Sociology, Social Anthropology and Mass Communication borrowed the meaning of function primarily from the biological sciences where it referred to the 'vital organic process in which they contribute to the maintenance of the organism'.

Stinchcombe (1968) writes: "In a functional analysis approach, researchers consider the consequence of some behaviour or social arrangements as an essential element of the causes of that behaviour". To put it simply, the roots of functionalism go back to the stimulus-response theory.

Thus the concepts of structural and functional linkage explain how tne different sub-systems are linked together to form the whole—the system of transfer of technology and the types of functions performed by each sub-system in order to make the process of transfer of technology effective.

Based on these concepts there were many studies pertaining to the TOT of improved agricultural technology. These studies culminated in identifying three distinct systems involved in the TOT of any improved agricultural technology viz. (i) Research System responsible for developing the innovations, (ii) Extension System responsible for disseminating the innovations, and (iii) Client System who puts into practice the disseminated technology. It also needs to be pointed out that all the three systems interact with each other constantly in the transfer of technology process. The authors of these studies include Axinn and Thorat (1972), Rao (1972), Singh (1974), Ambastha and Singh (1976) and Khan (1980).

Jaiswal and Arya (1981) expressed the view that the four sets of activities involved in the task of transfer of technology were: (a) research/technology production system which evolves technology; (b) extension/dissemination system which communicates the technology; (c) technology utilisation system/client system which utilises the technology; and (d) support system/input system which supports the transfer processes by providing facilities for processing and marketing of outputs.

Reddy (1981) pointed out that transfer of technology was a highly complex process involving several macro and micro level factors which he examined at length under seven systems viz. research, extension, client, input, economic, psycho-socio-cultural and administrative systems.

Sharma (1982) explained the transfer of technology process involving five sub-systems viz. research, input, extension, credit and client systems. According to him, farmer is the hub-the focus of the other four sub-systems to adopt technology appropriate and beneficial to him.

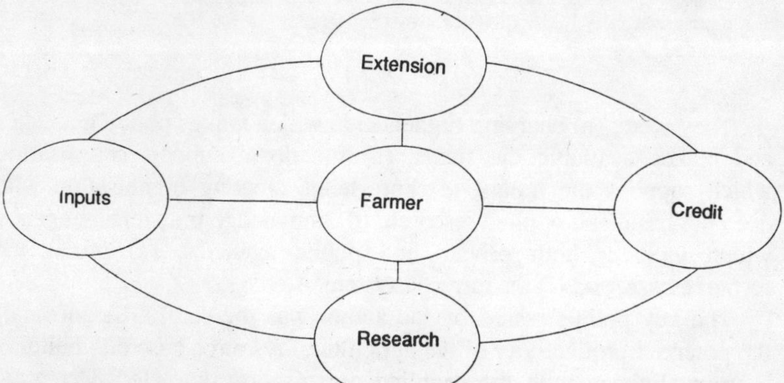

Research is shown as the base — the fountain head of scientific findings constantly renewing technology and evolving solutions to the problems.

Inputs and credit are the side support in the framework of agricultural production without which the best of technology would prove barren, abortive and merely abstract.

Extension organisation extends scientific findings to the farmers for use/for progress or economic gain.

Sen (1984) identified four systems involved in the system of transfer of technology. They were research, extension, client and input systems.

Lowdermilk (1985) postulated six basic system components for developing more effective extension systems depicted as follows:

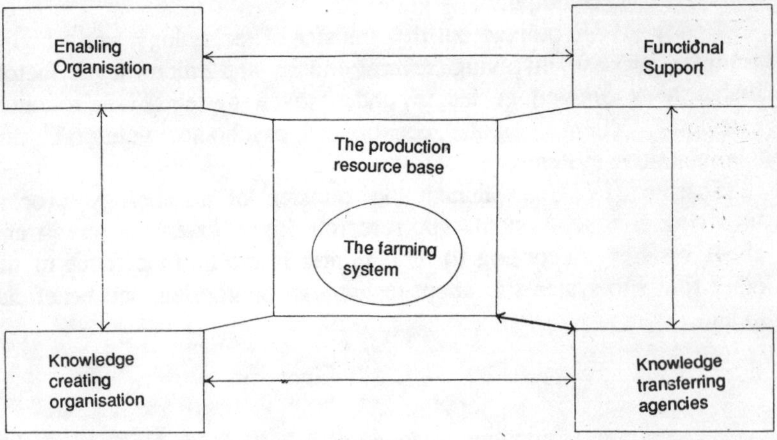

They were: (a) enabling organisation which makes policy mandates and makes available the funds, (b) functional support organisation which supplies the inputs, (c) knowledge creating organisation who does the basic and applied research, (d) knowledge transferring agency which includes both private and public agencies, (e) production resource base, and (f) the farming system.

The key point stressed by the author was the careful diagnosis of the potential productivity of the agricultural resource base and building a strong linkage with the enabling organisation, knowledge creating and knowledge transferring organisations. His model suggests a strong linkage between production resource base and knowledge transferring system.

Ambastha (1986) conceptualised three systems as involved in the communication process. They were (a) innovation development system, (b) extension system, and (c) client system.

Kishore (1986) suggested four systems for effective transfer of technology. They were (a) knowledge generating system (KGS); (b) knowledge disseminating system (KDS); (c) Knowledge consuming system (KCS); and (d) Input supplying agency (ISA). The KGS includes all the organisations involved in agricultural research including Agricultural Universities, ICAR institutions and research and production units of private and public sectors. The KDS is composed of all organisations engaged in dissemination of knowledge. Farmers, rural artisans and members of input agencies will constitute the KCS.

ISA includes seeds, fertilizers, pesticides and credit distributing agencies. Suggesting an alternate strategy for transfer of technology, Kishore developed the following model:

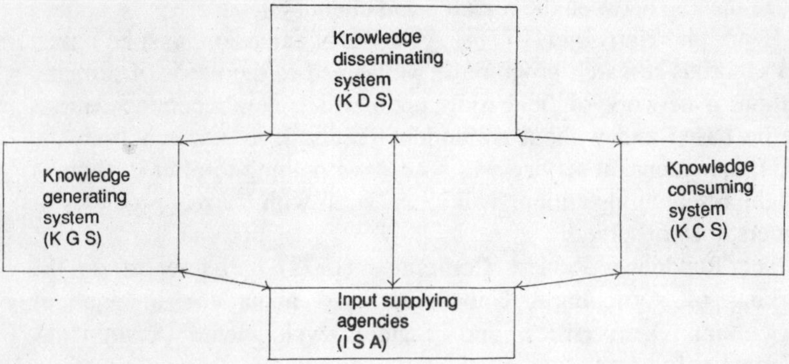

Bhatnagar and Desai (1987) opined that the overall system of extension and transfer of technology depends on effective and integrated functioning of the following:

a) Research management system
b) Extension management system
c) Support and sustenance management system and
d) Target beneficiaries management system.

The performance effect and impact of any technology delivery have to be measured in terms of their ultimate speedy communication, adoption and diffusion.

From the foregoing discussions it could be concluded that the transfer of technology system consists of four structurally and functionally linked sub-systems viz. (i) Research sub-system (ii) Extension sub-system (iii) Client sub-system and (iv) Input sub-system. The linkages between and among these sub-systems are of strategic importance in the process of transfer of technology of improved rice varieties released by the KAU.

2.3 Importance of Structural and Functional Linkages

The National Commission on Agriculture (1976), in its report considered it necessary for Agricultural Universities and concerned State Departments to co-ordinate their efforts in order to develop useful technology and effectively serve the farming community.

Allen (1977) stressed the need for some kind of linking system to bridge the gap between the research and client systems.

Benor and Harrison (1977) suggested that extension must be linked to a vigorous research programme well tuned to the needs of farmers. Without a network of field trials upon which new recommendations can be based and without continuous feedback to research from the fields, the extension service will soon have nothing to offer to farmers and the research institutions will loose touch with the real problems of farmers, they observed.

The Randhawa Review Committee (1978) in its report on the working of Agricultural Universities and about the linkages of Agricultural Universities and State Development Departments, observed as follows:

> "We would however like to mention that the linkage arrangements can only provide the formal framework of coordination. It is essential that the Agricultural Universities and State Government should devote considerable time and effort to give practical shape and content to such memoranda of understanding. For this purpose it is imperative on the part of senior officers and scientists to make a deliberate attempt to develop harmonious working relationship through both formal and informal contacts".

Jaiswal and Arya (1981) found that there was usually no effective link between the research and extension systems, a situation detrimental to the effectiveness of both. Without a continuous flow of innovation, the extension service would rapidly run out of anything to transfer. The authors also reported that a study conducted by National Institute of Rural Development (NIRD) in Rajasthan and Madhya Pradesh brought to light that there was lack of effective linkage between the research stations and the SMS which was one of the greatest weakness of T & V System in operation.

Reddy (1981) suggesting seven systems in the chain of technology transfer said that all the seven systems viz. research, extension, client, input, economic, psycho-socio-cultural and administrative and organisational systems are inter-related so much as to be deemed as the essential links in the chain of technology transfer. None of these interlinked systems should be underrated, much less ignored because the chain is as strong as its weakest link.

Pandey and Mishra (1984) in a study on communication behaviour of SMS in the T & V System found that systematic linkage of SMS with various sources of farm information provides the T & V system its initial start-up force and was 'a basic requirement for proper processing and transmission of relevant technology to the extension workers and farmers.

The study also revealed that the SMS lacked contact with information input sources and their output and processing functions were also poor.

Sen (1984) stated that for effective transfer of technology, strong inter-organisational linkage is of vital significance because of the involvement of various organisations in the process.

Swaminathan (1988) opined that the country has made considerable progress in developing research system, educational system and reasonably efficient input production and delivery system.

Venkataraman (1988) stated that besides extension, other systems like research, input supply, credit, infrastructure, markets and socio-economic environment have a great influence on adoption of improved agricultural technology. The author also observed that extension service should strive for effective technology delivery and promotion of adoption and establish functional linkages with related support systems.

From the aforesaid observations and discussions, it could be concluded that TOT is a complex and complicated task involving research, extension, client and input sub-systems. Ultimate impact of the new technology will depend upon the efficiency of the prevailing sub-systems and meaningful or effective linkage among these sub-systems. These facts bring into focus the need for and importance of effective linkage between and among the different sub-systems.

2.4 Concept of Transfer of Technology

2.4.1 *Transfer*

The term 'transfer' was defined differently. According to Chatterjee (1974), transfer means some sort of change of position, usually from one location or system to another location or system. The second definition given to 'transfer' by Chatterjee was in terms of an interaction between learning or performance across different tasks, varying across one or more dimensions.

The third definition given by him was more apt to the derived term 'transference' which has special relevance in the client-therapist interactional situation dealt in psychological literature.

Samantha and Kishore (1984) said that transfer may be conceived as equivalent to teaching. Transfer is therefore defined as the successful creation of opportunities or situations in which people gain the abilities and the situation necessary for successfully meeting their needs and interests in such a way as to attain continuous improvement and self-satisfaction.

The authors further argued that the task of change agents was to keep in mind not only the transfer of knowledge and skills but also fruitful application of these knowledge which include technological development, application, marketing and management of these technologies.

2.4.2 *Technology*

Schumacher (1973) opined that any technology developed to the scale appropriate to the human needs and satisfying to them may be called as appropriate technology.

Koontz *et al.* (1980) defined technology as the sum total of knowledge of ways of doing things. It includes inventions, techniques and the vast store of organised knowledge of how to do things.

According to Rogers (1983), technology is a design for instrumental action that reduces the uncertainty in the cause-effect relationship involved in achieving a desired outcome. A technology has two components (a) hardware aspect consisting of the tool that embodies the technology as material or physical objects and (b) software aspects, consisting of the information base for the tool.

A technology cluster, again, according to Rogers, consists one or more distinguishable elements of technology that are perceived as closely interrelated. Some change agencies promote a cluster of package of innovations because they found that the innovations are thus adopted more rapidly.

Thus technology is a means of uncertainty reduction for individuals that is made possible by the information about causa — effect relationship on which the technology is based.

2.4.3 *Transfer of Technology (TOT)*

Dwarakinath and Channegowda (1974) pointed out three transfer deficiencies. Firstly, not all the available technology is transmitted to the field. It appears that only those elements that make a conspicuous impact on application are taken care of. Secondly, not all potential adopters get exposed to the new information to the same degree. Thirdly, even among the adopters of new technology all the elements of technology are not adopted.

Verma (1974) defined TOT as shifting a technology from one person to another and from one place to another place.

Jaiswal and Arya (1981) defined TOT as a process by which the recommended practices produced by research and development agencies are transmitted through extension agents to producers. Therefore, TOT starts after its perfection and ends in its utilisation by the target consumers.

Reddy (1981) expressed the view that effective TOT takes place when the maximum number of potential adopters understand, accept and actually put into practice the major part of an item of technology with the minimum time lag and with the maximum possible material and financial benefits.

2.4.4 *Linkages for TOT*

The National Seminar on Challenges of Extension in Eighties (1982) made certain recommendations for streamlining the existing linkages among research, extension, farmers and various other infrastructures connected with rural and agricultural development programmes. Some of the major recommendations included the following:

a) Linkages should be established wherever they do not exist and existing linkages need be strengthened. The linkages to be established or already existing should be institutionalised; built-in mechanism for effective collaboration should be provided and the linkages should be activised.
b) Linkages should be evolved at as many levels as are operative and pertinent.
c) Linkages have to be vertical as well as horizontal. Presently vertical linkages are working to some extent. So horizontal linkages need to be strengthened.

d) Extension advisory committee should be constituted in all Agricultural Universities. Representation in such committees for farmers, farm labourers, extension workers, input organisations etc. was found weak and therefore should be strengthened.

e) System of monitoring at all levels should be created for keeping the linkages revitalised and effective.

f) The frequency of meeting of co-ordination committees for the linkages has to be much higher at the lower levels.

g) *Kisan melas*, workshops, etc. should be carried out at district and block level than at State level only.

h) The linkages between research and extension will be further strengthened if the SMS are persons specialised in the subject at post-graduate level.

i) An inventory of the technology evolved by individual farmers and any other organisations should be prepared and all such techniques should be verified. This will strengthen the linkages between researchers and client systems.

j) Linkages established between extension and research through T & V system should be further strengthened.

Sadikin (1985) was of the opinion that there are four major channels for technology transfer (a) Extension, which comprises, technology generation, verification through on-farm demonstrations, extension assistance and application by farmers; (b) The direct transfer of technology from National Agricultural Research Systems to agricultural farms and to progressive farmers who have the capability to adopt technology to local needs; (c) Research results are passed through policy and decision makers at national and regional levels and (d) Research results are disseminated through Universities and Agricultural Schools.

Gee (1981) has pointed out that new technology to be effective, managers must be sensitive to environment, there must be an understanding of the users' needs, effective communication channels should exist, individuals committed to expediting the new technology must be present and there must be an organisation with wide identity and support for the new technology.

Prasad (1984) found that the question of near stagnation of production is not due to technological reasons, rather it is due to slow transfer of technology. He further stated that transfer of agricultural technology is a complex process involving research institutions, educational, training and marketing agencies and farmers.

Samantha and Kishore (1984) opined that most effective system for diffusion and adoption of technology depends upon efficient functioning of four systems viz. research, extension, client and support systems.

Singh (1984) viewed that the success of TOT will depend upon understanding the nature and characteristics of the technology and their specific requirements, the characteristics of the farming community and effective communication strategy.

Okuneye (1985) suggested a group approach involving demonstrating new technology to farmers on their farms for rapid transfer and adoption of improved practices.

A perusal of the literature reviewed on linkages for TOT brings to focus the need for effective linkages between and among the four sub-systems viz. research, extension, client and support systems. The recommendations of the National Seminar on Challenges of Eighties also reinforced the need. However, it has to be noted that linkages are not totally lacking in India, nonetheless they require refurbishing on sound footage.

2.5 Concept of Systems Analysis

The *Webster's Dictionary* defines a system as "a regularly interacting or independent group of items forming a unified whole".

A system has been defined in the *Oxford English Dictionary* as "a set or assemblage of things connected or interdependent and interacting so as to perform a complex unity, a whole composed of parts in orderly arrangement according to some scheme or plan". For any system, there must be boundaries that separate it from its environment.

Hall and Fagen (1956) defined system as a set of objects together with relationships between the objects and between their attributes.

Buckley (1964) defined a system as a complex of elements or components directly or indirectly related in a causal network, such that each component is related to at least some others in a more or less stable way within any particular time.

All the above definitions epitomise the interdependence and interacting nature of different components which constitute a system.

Johnson *et al.* (1967) opined that systems analysis provides the means for considerations and integration of all functions necessary for successful mission accomplishment and emphasises the optimisation of total system performances. It also provides for evaluating alternative means of mission accomplishments in terms of cost and effectiveness.

Quade and Boucher (1968) expressed the view that systems analysis is a research strategy, a perspective on the proper use of the available tools, a practical philosophy of how best to aid a decision-maker with complex choices under uncertainty. They further added that systems analysis was a systematic approach of helping the decision-maker choose a course of action by investigating his full problems, searching out objectives and alternatives and comparing them in the light of their consequences using an appropriate framework.

Chand (1971) defining research utilisation said that research utilisation was not a one-way flow of research results from researchers to clients, but also entailed a flow of client needs to researchers. The processes involve the reflection of the needs of clients to researchers through linkers, researchers complete research on client centred problems and finally the clients utilise the research results. In this process the three social systems involved are the research, the linking and the client systems.

Dubey (1974) suggested a "whole system" analysis to study the adoption-proneness of farmers.

According to Singh *et al.* (1975) a system is characterised by wholeness, non-summativity, feedback and nature of finality.

Wholeness implies that every part is interrelated, that a change in one aspect will cause a change in all of them. Non-summativity means that a system is not the sum total of all its parts. Feedback implies that a system is not characterised by casual relationship but by circular and complex relationship. The nature of finality indicates that the future behaviour of a system is not determined by the initial conditions but by the nature of the process or by system parameters.

Based on the degree to which a system interacts with its environment, the authors classify system into two types: (a) open system and (b) closed system. A system is open not simply because it engages in interchanges with environment, but because this interchange is essential to the systems viability, continuity and its ability to change. The typical response of a natural closed system to an intrusion of environmental events is loss of organisation, or change towards dissolution of the system.

Baird (1977) said that a system is a group or combination of component parts arranged in such a way as to constitute a unified whole.

The system school considers a flow originating in the environment that moves into and through the organisation and feedback into the environment. Involved in this flow are information materials, people and behaviour. In other words, that which the system receives from the environment is input. Energy and information form the input for organisations. That which the system places back into the environment is considered as output. Two types of output can be identified – products and services.

Sampath *et al.* (1981) stated that systems analysis essentially is a systematic way of identifying goals of any system and scientifically working out different steps to move towards these goals, suggesting models for application. The authors further state that systems approach is a rational problem-solving method of analysing the educational process and making it more effective.

According to Mann (1983), a system analysis specifies the set of principles that govern the interconnections of the system parts.

Ambastha (1986) considers that the concept of system analysis is an outgrowth of second world war operation research and no precise and commonly accepted definition exists so far. The essence of systems analysis is to construct and operate within a model – an idealisation of the situation; appropriate to the problem. Such a model which may be either very simple or very complex introduces a precise structure and terminology that serve primarily, as an effective means of communication enabling the participants in the study to make their judgement in a concrete context. Moreover, through feedback, the model helps the decision-maker, the analyst or the expert on whom they depend to arrive at a clear understanding of the subject-matter and the problem.

The whole act of systems analysis can be broadly subdivided into five activities (Quade and Boucher, 1968). They are: (i) formulation stage: the issues are clarified, the extent of enquiry is limited and the elements are indentified; (ii) search stage: information is gathered and alternatives generated ; (iii) evaluation or analytical stage: a model or set of models are built with alternatives and alternatives are compared in terms of their consequences; (iv) interpretation or judgement stage: predictions from the models are obtained, conclusions are derived and the causes of action are indicated; and (v) verification stage: the conclusions are tested by experiment.

2.6 Studies on Linkages for TOT

2.6.1 *Linkage between Research Sub-system and Extension Sub-system*

Kelsey and Hearne (1963) recommended that research and extension systems should make available quickly to the rural people the results of research in usable form that are applicable to and will benefit farmers and their families.

Moseman (1970) recommended the following linking functions for research workers:

a) Organising teaching and training programmes at selected research centres designed to cater to the needs of specific agricultural specialists.
b) Establishing close linkage between the research centre and extension and other activities responsible for helping farmer to use new technology.

Havelock *et al.* (1969) stated that the researchers himself may be only a competent conveyor of many kinds of research results. The researchers and development personnel should translate research into usable service and products.

Mirdha (1974) said that the process of modernisation of agriculture depends upon the extent of integration between the research system and extension system and also to the extent to which those engaged in these two functions have mutual understanding and cooperation at various levels.

Mosher (1975) expressed that fostering linkage with complementary organisations is a strategic task. It can be done by taking a genuine interest in the work of other organisations, acknowledging the mutual dependence that exists and getting personally acquainted with administrators of other programmes. Any contact with them is good, but going to them is far superior to calling them in.

Reddy (1981) suggested that the scientists should guide extension workers and use the feedback from them.

Singh (1974) summarising the decisions of the symposium on transfer technology-systems and constraints suggested that there was a need for better representation and greater involvement of extension personnel, farmers, input supply and credit agencies in various research committees at various levels.

Jaiswal and Das (1981) suggested that effective transfer of appropriate technology to rural people ideally involves establishing appropriate system for ensuring proper linkage between research and extension systems. They also observed that a farmer's decision to accept or reject an innovation depends not only on personal characteristics but also on the effectiveness of various agencies engaged in this process. The first step in the process of transfer of technology is not the development of new ideas or technology by the scientists, rather it is communication of existing levels of technology use and problems of the farmers to scientists in order to facilitate the development of relevant and need based technology. Lack of link between extension and research systems is detrimental to both. In India the link between research system and grass root level extension worker is by and large through farm journals, magazines and extension literature, the authors further added.

Surendran (1982) in a study on impact of operational research project in Kerala found that one of the main problems in the TOT was lack of co-ordination between the Agricultural university and the State Department of Agriculture.

Eklund (1985) indicated that adoption of new technology in peasant agriculture remains lower than expected. Failure to achieve expected returns associated more with insufficiencies in the conduct of research than with external and economic factors such as price structure or lack of agricultural support services. The author emphasised the need for on-farm research with system approach.

Benor and Baxter (1984) suggested field visits as important means of research-extension linkage. Research staff must make frequent field visits to review specific field problems raised by extension staff. The authors further suggested that just as research support is required for effective extension, extension staff have an important role in supporting research. Frequent visits to the field and participation in joint meetings, workshops and farm trials are vital for the necessary mutual support of extension and research and every possible step should be taken by extension and research staff to establish and strengthen the formal and informal linkages upon which this support depends.

The National Seminar on Professional Agricultural Extension System (1984) emphasised the need for closer link between research and extension agencies.

Singh (1984) pointed out that the linkages between extension and research systems are still very weak and of *ad hoc* type. A stage has reached that a regular mechanism has to be established for satisfactory communication between these two systems. Without continuous and free flow of crop production technology from research institutes to farmers' fields the extension service becomes non-professional and looses its credibility and has nothing new to extend.

Sabarathnam (1985) in a study conducted among 32 rice and sorghum scientists belonging to Agricultural Research Service, found that a considerable percentage of the scientists did not perceive that they should have dissemination linkage with extension workers, farmers and input agencies, they do not think that they should have dissemination linkage with the organisations other than those dealing with inputs. Most of the scientists maintain dissemination linkage through training lectures, office call of farmers and extension workers, journals, discussions with state department officials, advisory letters and demonstrations.

Bhatnagar *et al.* (1986) in a study on Management of agricultural extension under T&V system in different States in India found that majority of the States in India indicated satisfaction regarding research-extension linkages.

Desai (1987) stated that research and extension are the two basic crucial functions in agricultural development and research and extension activities are highly interdependent. Extension to be effective requires research findings to reach farmers. Similarly, research to be more effective requires constant feedback of the issues and problems of farmers on which researchers should focus attention.

Pickering (1987) found that there was a need for closer liaison between research and extension to ensure that research programme are relevant to farmers' needs. This requires among other things, that validation programmes are effectively and jointly carried out on farmers' fields with active farmer participation. To build up effective linkage between research and extension the author suggested the following methods:

i) There must be increased emphasis upon on-farm research with special attention to socio-economic studies and feedback from extension staff and farmers.
ii) There must be a greater inter-disciplinary collaboration between and among researchers, farmers and extension staff working in

the field.

iii) All agricultural institutions, policies and procedures must become more responsive to the interests and needs of the farming population particularly to the vast majority of the small farmers.

iv) Whether extension or research, management must be alert, dynamic, responsive to local and national needs.

Malekmohammadi (1987) found that researchers in agriculture need creative relations with extension workers and also agricultural educators and *vice versa* in order to activate extension movement and at the same time research and education activities in agriculture.

Venkataraman (1988) stated that after the introduction of T&V system, there was a well-knit extension-research linkages through a team of subject-matter specialists in the country. But both the extension and research systems should learn to work as a single unit for rendering effective service to the farmers. The author further added that the success in agricultural development can come about only through their combined efforts and hence it is essential to foster continuously strong extension and research linkages.

Anantrao (1988) stated that the first priority in India after independence was to have the agricultural education programmes based on three elements-research and development of appropriate technology, extension of technology to practising farmers and university level training to have necessary man power for research and service. An organic linkage among these three elements was needed to ensure that there was a two-way flow from farmers to the research laboratory and class rooms and *vice versa*.

A critical review of all the available literature on research-extension linkages emphasises the need for a close-knit relation – just like two sides of the same coin – between research and extension. Although the research system is mainly concerned with the creation of new technology. It is not an end in itself. The ultimate object is the adoption of the technology by the clientele. Thus the research system can link itself with extension and client systems by devoting part of their time for extension activities. The extension system which is considered to be the king-pin in the entire mechanism of agricultural development should also play an equal role in the process of TOT of improved rice varieties.

2.6.2 *Linkage between Research Sub-system and Client Sub-system*

Beal and Rogers (1959) reported that farmers generally viewed research workers as distant referent and some one with whom they had little direct contact.

Havelock *et al.* (1969) stated that central to the problem of information production, dissemination and use is the question of linking roles. The need for a link between the knowledge production and user system is that scientists and farmers are the products of two distinct worlds.

Kamath (1974) observed that there was a time when a scientist's function ended when he wrote up the findings of a research project and had published it in a scientific journal to be read by his counterpart. But present day field-based and problem-oriented research gave him necessarily the function of translating the results for consumption by farmers and extension specialists.

Ambastha (1977) in a study of communication pattern of farm scientists in Bihar State found that researchers in general had comparatively higher contact with farmers visiting office, influential farmers and farmers visiting kisan melas. But traditional farmers were the least contacted categories.

Dudhani (1980) in a study on the linking roles of extension supervisors subject matter specialists and research workers in Karnataka found that research workers performed the following functions in a poor way:

a) Discussing the programme of adaptive trials with farmers during the field days.
b) Writing newspaper articles relating to recommendations made from the research stations.
c) Visiting neighbouring villages periodically.

Jaiswal and Das (1981) in a study on TOT of rice farming found that effective TOT to rural people ideally involves the following activities:

a) On specific subject areas, understanding the present level of usage of technology in different regions/areas and identify the types of technology needed to meet the problems of rural people.

b) Based on such feedback, generation of appropriate technology in the field of agriculture, and other needed and pertinent problems in rural life.
c) Appraisal of the technology by conducting required field trials under different agro-climatic and socio-economic conditions.
d) Organising extension and educational efforts for the diffusion of the technology in the rural social system.

The authors also observed that lack of communication between farmers and research scientists resulted in developing publication oriented research results rather than farmer or field conditions oriented results.

Nweke and Akorhe (1985) based on case study on rice production in the plateau of Nigeria, reported that a technology is adopted or not depends upon whether or not it is profitable in the production activity. For wide adoption, a technology should be properly tested by researchers for profitability in terms of the farmers' needs and circumstances.

Kaur (1985) in a case study on TOT of smokeless chullas at Wazirpur village in Haryana found that while giving inputs of new technology, the attitudes, interests, perceptions, inhibitions and restrictions of villagers are not taken into account. Often new technology is developed from the scientists' point of view only. What is most needed is the creation of a rapport between scientists and the villagers.

Prasad and Reddy (1986) stated that the approach of farming system research rests on two central positions:

a) that effective research on agricultural technology starts and ends with farmer; and,
b) that integration of perceptions of biological scientists and social scientists is an essential element in such a research.

The reviews highlight the fact that if a technology is to be accepted by the target farmers, it should be need-based and problem-oriented for which an organic linkage between the researchers and clients is undoubtedly needed. The initiative for this should come from the scientists as, for vast majority of small and marginal farmers, the scientists are inaccessible.

2.6.3 *Linkage between Research Sub-system and Input Sub-system*

Jaiswal and Arya (1981), Reddy (1981), Sharma (1982), Sen (1984), Kishore (1986) and Bhatnagar and Desai (1987) emphasised the need for effective linkage of input system with research, extension and client systems for effective TOT.

Singh (1984) stressed the need to understand the process of farm technology development, communication and integration. This requires many and repeated behavioural changes not only among farmers, but also among researchers, extension workers, communicators and distributors of inputs and commodities.

Swaminathan (1988) expressed the view that research on knowledge and input delivery systems which can help all farmers irrespective of their innate input mobilising and risk taking capacity is needed to derive economic benefit from new technologies. Very little attention has been paid so far to research on delivery system, he observed.

2.6.4 *Linkage between Extension Sub-system and Client Sub-system*

Edward (1957) recommended to strengthen the linkage between extension specialists and farmers by keeping constant contact between the two. Also he recommended to use simple language while writing research publications.

Dwarakinath and Channegowda (1974) in a study conducted in Karnataka found that the linkage between extension system and client system has not been developed thoughtfully and systematically.

Awa and Crowder (1978) stated that extension agents are not the most frequently contacted source, but they do seem important as "information validators". Farmers know that they can depend on extension for timely information about new farming developments. In general, the findings indicate a tendency for farmers to look to other sources for initial information with the extension agent assuming an intermediate role. The authors concluded that extension agents should assume an active role in which they function as one of many communication links.

Dudhani (1980) in a study conducted in Karnataka found that none of the linking roles formulated according to job chart of Assistant Directors of Agriculture with clients, were perceived in a highly satisfactory way, while the Deputy Directors of Agriculture recognised the importance of various specific linking role functions in an adequate way.

Jaiswal and Das (1981) stated that extension programmes lack long-term educational approach. Extension programmes are more programme oriented than farmer oriented and that demonstrations are not conducted properly. The authors also observed that VLWs' visit to farmers depended much more on the VLWs personal motivation and interest. Effective transfer of appropriate technology to rural people involves organising extension and educational efforts for the diffusion of technology in the rural social system.

Rogers (1983) stated that change agents provide communication link between a resource system of some kind and a client system. One of the main roles of the change agent is to facilitate the flow of innovations from a change agency to an audience of clients. But this type of communication to be effective, the innovations must match the needs of the clients. And for linkage to be effective, feedback from the client system must flow through the change agent to the change agency so that it can make appropriate adjustments on the basis of previous successes and failures. He also suggested a sequence of roles to be performed by the change agents. They were:

a) develop need of change among clients
b) establish an information-exchange relationship
c) diagnose the clients' problem
d) create desire for change in the client
e) translate interest into action, and
f) achieve a terminal relationship between extension and client systems.

Singh and Babu (1984) in a study conducted at Indian Agricultural Research Institute indicated tremendous gap in communication of technology at various levels, especially between extension personnel and farmers.

The National seminar on Professional Agricultural Extension System (1984) *inter alia* emphasised the need for closer link with farmers and programme implementing agencies for augmenting agricultural production.

Sethy *et al.* (1984) in a study on interpersonal characteristics in adoption of farm technology highlighted the impact of entrepreneurial characteristics of the farmers on their adoption of farm technology.

Technological knowledge on farm can be passed on to the farmers of all ages without any apprehension about their learning ability. Personal as well as influence motivation of the farmer boosts up the gain of technological know-how by them which is the key variable to effect adoption of HYV rice technology.

Lowdermilk (1985) is of the opinion that strong functional linkage with support and service organisations, policy making levels, knowledge creation centres as well as other private and public agencies involved in the transfer of knowledge is necessary to effectively provide new and improved production possibilities to farmers for the following reasons:

a) Agriculture is inherently complex due to the biological nature and great variability.
b) The type of agriculture practised in an area is largely determined by the existing physical resource base, the prevailing market demand and local traditions.
c) Any extension system should also be based on a careful diagnosis of the particular stage of agricultural development prevailing in an area, and
d) The great variability in the agrarian structure.

Reddy (1986) in a study on extension systems interaction with research and client systems in Andhra Pradesh concluded that the potential sources of communicative interactions like field days, exhibitions, cattle fairs, tours, film shows and publications should frequently provide the platform to extension workers for communication purposes. The poor access to most of these media appear to be the hindrance in the process of technological diffusion.

Venkataraman (1988) suggested that research system and extension system should learn to work as a single unit for rendering effective information source to the farmer.

The above reviews bring into focus the need for building a closer linkage between extension system and client system which form two major links in the chain of TOT. The prime functions of extension system is to translate the research results in a way digestible to the clients and the client system in turn should feedback its observations and results to the extension system to be passed on to the research system.

2.6.5 *Linkage between Extension Sub-system and Input Sub-system*

Singh and Prasad (1970) in a study on coordination between agriculture and cooperative department at state, block and village level revealed that there was neither very good nor even good coordination at any of the levels. The authors suggested that for desirable coordinated system of administration, the planners may have to control the human as well as non-human weaknesses that are likely to tell upon the efficiency of the personnel.

Jaiswal and Arya (1981) in a study on TOT in rice farming concluded that effective transfer of appropriate technology to rural people ideally involves creation of necessary infrastructure for accelerating the input supplies including credit as also the management of the output resulting from the use of such technology.

Venkataraman (1988) expressed the view that it is desirable the farmers come to see extension as a 'reliable knowledge source' and not as a 'dispenser of subsidised inputs'. He suggested that input use can be promoted through production performance rewards, rather than by subsidised sale.

The foregoing studies indicate that in the system of TOT of agriculture, input is a potential sub-system and its functional linkage with all the other sub-systems viz. research, extension and client sub-systems should be ensured.

2.6.6 *Linkage between Client Sub-system and Input Sub-systems*

Jaiswal *et al.* (1970) found that small farmers were not slow to change and they can keep pace with big farmers in the matter of adoption if arrangements were made for the supply of inputs. This calls for a better co-ordination of farmers and input supplying agencies.

Coombs and Ahamed (1974) considered that the deficiency of ineffective delivery system to small farmers is primarily due to lack of management service. Management service is required to assist the small farmers in getting supplies, services, credit, help in marketing etc.

According to Awe and Crowder (1978) commercial dealers were found to be readily accessible source and were in a strategic position to communicate with farmers when their credibility has been established. Clearly they are determining force in decisions to adopt new farm practices. However, when a final decision must be made, it is fellow farmers who are most often consulted.

Amalraj and Prasad (1984) in a study on role of fertilizer dealers in the transfer of new farm technology in Rajasthan found that fertilizer-pesticide dealer was one of the important sources of information, as 16.80 per cent farmers utilised this source.

Matlock (1985) opined that input system was the side support in the framework of the agricultural production phenomenon. These organisations make and provide the inputs without which the best technology would prove barren, abortive and abstract.

Ensminger (1986) expressed the view that if the small farmers is to significantly improve his yields, he must be aided with agricultural technology within his resources and competence.

Input sub-system is an easily accessible source of information to the farmers. But often the farmers look upon the input supplying system as a profit making concern and perhaps this fact keeps the client sub-system and input sub-systems far apart. This situation warrants concerted effort on the part of client, extension, research and input sub-systems to build a strong linkage between and among themselves.

2.7 Constraints in the TOT of Improved Rice Varieties

According to Mann (1983) most economic decision-making involves choosing an optimum strategy subject to the limits imposed on the decision maker. These limits are often referred to as constraints.

The Programme Evaluation Organisation (1969), in its evaluation report of the HYV programme, reported the following constraints in the adoption of HYVs of rice and jowar: (a) Attack of pests and diseases (b) Lack of marketing facilities (c) Untimely supply of fertilizers (d) Non-availability of plant protection chemicals (e) Lack of irrigation facilities (f) Problem of easy and adequate credit (g) Labour shortage and high cost of labour.

Joon *et al.* (1970) stated that in order to popularise HYVs, there is a need to lay greater stress on educating the farmers about the finer technicalities of latest varieties. Intensification of extension activities, provision of crop loan facilities and releasing varieties with better grain qualities are factors for making HYV popular.

Garg and Omprakash (1974) felt that heavy investment needs, non-compatibility with consumer needs and complexity of modern technology are the most important barriers coming in the way of production technology of HYV of crops in the farmers' fields.

Pal (1975) reported that major constraints to paddy yield were related to the difficulties in obtaining seeds, chemicals, fertilizers and credit, inadequate irrigation water and attack of pests and diseases.

Sharma and Nair (1974) in a study on HYV paddy conducted in Kerala found that lack of irrigation facilities, high incidence of pests and diseases, lack of conviction about the profitability and lack of finance were the reasons for non-adoption of HYV of paddy.

Parthasarathy (1975) found that the most severe constraints to shifting to modern varieties in West Godavari district of AP were (a) difficulties in water management in the wet season, (b) bias of new technology in favour of large owners and (c) unfavourable position of the small owners in inputs, credit and markets.

Ray (1976) stated that lack of drainage facilities, incidence of pests and diseases and lack of suitable varieties were perceived as most important reasons for low spread of HYV of paddy during Kharif season in West Bengal.

International Rice Research Institute (IRRI) (1977) reported that successful adoption of the improved rice technology including modern varieties is a function of a number of factors like education, technical knowledge, hazard experience and extension exposure of farmers and availability of inputs, credit and perceived profitability of the varieties.

Jaiswal and Das (1981) reported that several attributes need to be considered carefully while developing a particular technology and transferring the same to the farming community particularly to the small and marginal farmers. These were as follows:

i) Attributes of the adopter (technical skill, attitude towards change and risk, level of aspiration and level of income).
ii) Aspects of technology (Profitability, divisibility, complexity, communicability and technical soundness of the technology).
iii) Characteristics of economy (infrastructure, demand for products, Government policy etc.).
iv) Features of environment at the farm level (socio-cultural factors, farm resource constraints).
v) Risk and uncertainty.

Tripathy *et al.* (1982) reported that constraints in adopting HYV of rice technology in coastal Orissa were: (i) difficulty in getting institutional credit, (ii) inputs neither received well in advance nor in adequate quantities, (iii) improved technology of HYV of rice cultivation has never been demonstrated to them, and (iv) non-availability of irrigation in May to prepare dry nursery.

Rajasekhar (1984) in an yield gap analysis in Madurai reported that 61.82 per cent of the farmers stated high cost of inputs as the main reason for the yield gap followed by lack of own funds, inadequate power supply and incidence of diseases.

Leagans (1985) opined that optimum adoption of agricultural production innovations is achieved only when a farmer is persuaded to accept a technical innovation as being for him technically sound, economically feasible, physically possible and politically and socially compatible.

Pachico and Ashby (1985) in a case study on adoption and screening of new rice varieties in Nepal found that rapid adoption of new technology by the small farmers requires firstly, decentralisation of biological screening to small scale site-specific trials with farmers and secondly, a focus of extension efforts on the evaluation of full scale management requirement with the participation of small farmers.

Singh and Sharma (1986) in a study conducted among 100 paddy growers in Kurukshetra brought out seven production constraints as perceived by the farmers in the order of their importance. They were:

a) High cost of agro-chemicals
b) High cost of fertilizers
c) High cost of HYV seeds
d) Non-availability of HYV seeds
e) Lack of adequate finance with farmers
f) Inadequate farm credit
g) Lack of irrigation facilities

The review of the constraints in the adoption of improved rice varieties indicates that some of the major problems like high cost of labour and other material inputs, high incidence of pests and diseases, lack of adequate credit and irrigation facilities and non-availability of improved varieties of paddy seeds suited to different agro-climatic and soil conditions are common to all rice growing states in India. In the light of these observations, it would be desirable to find out the major constraints as perceived by Kerala rice farmers.

2.8 Conceptual Model of the Study

The conceptual framework within which the present study was formulated is depicted in Fig. 2.1. Review of literature helped to uncover a number of studies related to the problem and the available techniques which could be used to study the structural and functional linkages in the TOT of improved rice varieties released by the KAU. Based on the literature reviewed, a conceptual model as shown in Fig. 2.1 was developed.

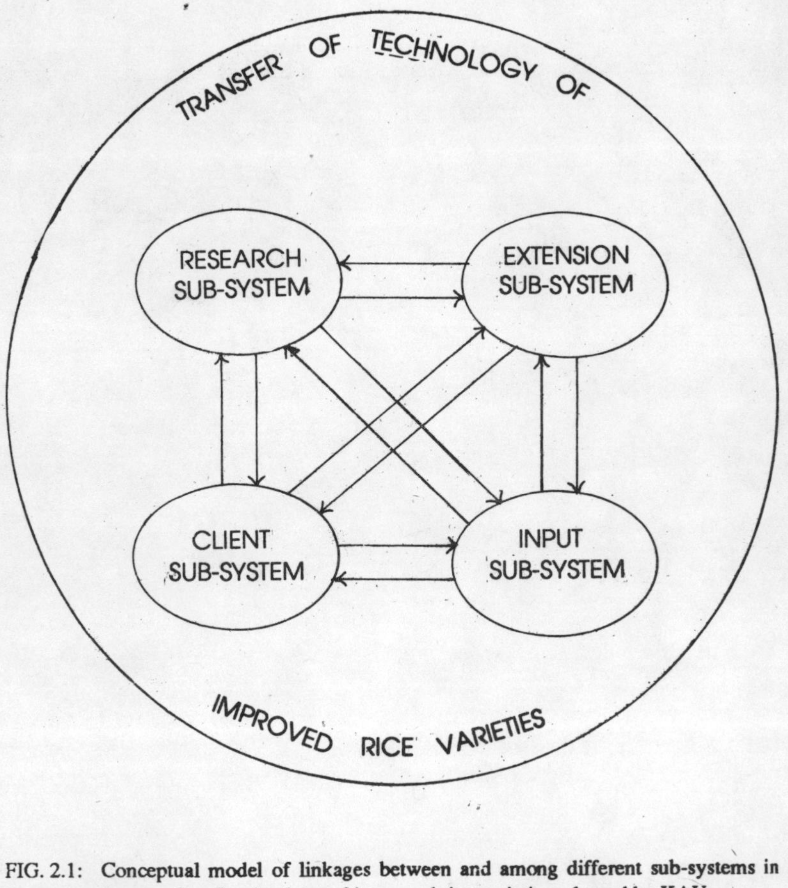

FIG. 2.1: Conceptual model of linkages between and among different sub-systems in the transfer of technology of improved rice varieties released by KAU.

It has been conceptualised that the TOT of improved rice varieties released by the KAU was the results of coordinated activities of four major sub-systems – Research, Extension, Client and Input sub-systems. The type of interaction expected between and among these four sub-systems is indicated by two-way arrows which symbolise a perfect structural and functional linkage between and among the four sub-systems in the TOT of improved rice varieties released by the KAU

3

METHODOLOGY

IN ACCORDANCE with the specific objectives, the methodology followed in the study is summarised under the following heads:

3.1 Identification of systems
3.2 Locale of the study
3.3 Selection of respondents
3.4 Selection of variables
3.5 Measurement of variables
3.6 Procedure adopted for data collection
3.7 Statistical tools used
3.8 Operationalisation of concepts
3.9 General hypotheses set for the study

3.1 Identification of Systems

Johnson *et al.* (1967) suggested that systems analysis would provide with the means for integration of all linkage activities.

Chand (1971) conceptualised that three systems viz. research, extension and client systems were involved in any research utilisation process.

Singh *et al.* (1975) stated that a system cannot function in isolation as each component of the system was related to atleast some other components in a more or less stable way at any particular time.

In a study on communication patterns of farm scientists, Ambastha and Singh (1976) followed a systems analysis approach conceptualising farm scientists of different cadres as sub-systems of the innovation development system.

Studies of similar nature had been conducted by Jaiswal and Arya (1981), Sen (1984) and Kishore (1986) and in all these studies the authors suggested that in any TOT process four systems were involved and they were the research, extension, client and input systems.

Though, the nomenclature differs slightly, the implied systems were the same in these works. In the light of the foregoing discussions and having set systems analysis as the research strategy in this study, it was felt such an approach will unfold the linkage activities between and among the different sub-systems in the TOT of improved rice varieties released by the KAU. Therefore, for the purpose of this study, it was conceptualised that the system of TOT process of improved rice varieties released by the KAU encompasses four major sub-systems viz. (i) Research sub-system (RSS); (ii) Extension sub-system (ESS); (iii) Client sub-system (CSS) and (iv) Input sub-system (ISS).

3.2 **Locale of the Study**

3.2.1 *Selection of the Study Area*

The KAU is responsible for formulating location-specific agricultural research programmes for the State and implementing them on a multi-disciplinary basis.

For this purpose, a research policy has been enunciated giving emphasis to solving the problems of farmers and increasing total agricultural production in the State as well as net income of the farmers. On the basis of the above policy, the research activities have been made more comprehensive to cover all economically important crops grown in the State. In the formulation of research projects, the extension personnel of the Department of Agriculture, progressive farmers and representatives of input agencies are also intimately associated through research advisory committee, seminars, workshops and others

The KAU after its inception in 1972, has evolved and released 14 improved rice varieties from its four major research stations, viz. Regional Agricultural Research Station (RARS) at Pattambi, and Rice Research Stations (RRS) at Vyttila, Moncompu and Kayamkulam. The present study envisages to find out the structural and functional linkages between and among the different sub-systems in the TOT of these varieties. Hence, the study was conducted in two of the agroclimatic regions of the State viz. the central region and the region of problem areas which form the service areas of the four research stations. The Agricultural Research Station (ARS) Mannuthy, though not directly involved in evolving a rice variety, is functioning as a sub-centre of the central region and region of problem areas. Hence,

ARS, Mannuthy was also included for the study.

The respondents of ESS, the CSS and the ISS were also drawn from the agricultural sub-divisions of the districts in which the RARS Pattambi, RRS Vyttila, Moncompu and Kayamkulam are situated. Thus, districts of Alleppey, Ernakulam, Palghat and ARS Mannuthy of Trichur district formed the locale of the study (Fig. 3.1).

3.3 Selection of Respondents

Since the study aimed at systems analysis of existing linkages for TOT of improved rice varieties, respondents from research, extension, client and input sub-systems were included for the study.

3.3.1 *Research Sub-system*

The RSS consisted of all the scientists of the KAU who had been associated with evolving and releasing improved rice varieties from the research stations at Pattambi, Vyttila, Moncompu and Kayamkulam. As the number of scientists working in these research stations was limited, a complete enumeration of the scientists was done for the study. Thus out of the 55 scientists in position only 52 of them were available at the time of data collection and they formed the respondents of the RSS. The number of scientists who responded from each research station is shown in Table 3.1.

TABLE 3.1: List of research stations and number of scientists responded

Name of research station		No. of scientists responded
RARS	Pattambi	19
RRS	Moncompu	18
RRS	Kayamkulam	7
RRS	Vyttila	4
ARS	Mannuthy	4
	Total	52

3.3.2 *Extension Sub-system*

The officers of the Department of Agriculture from sub-division to Agricultural Extension (AE) unit level were responsible for the TOT of improved rice varieties released by the KAU and these officials formed the ESS.

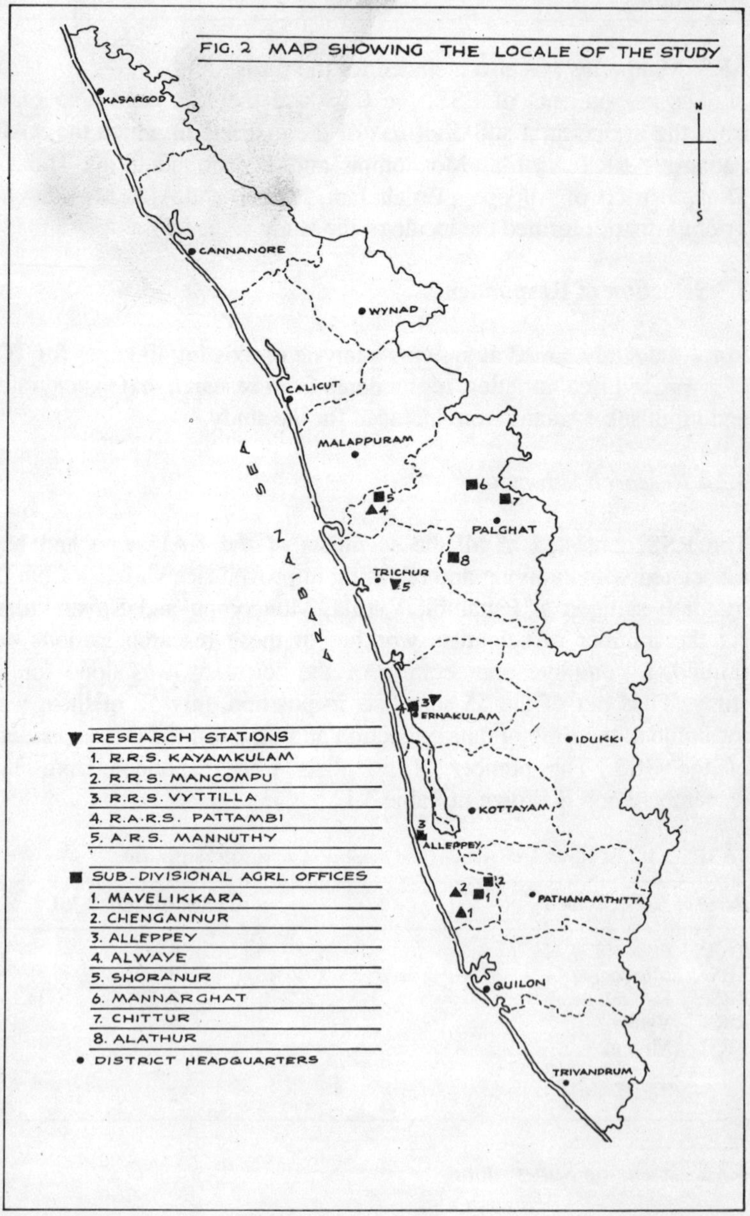

FIG. 3.1: Map showing the Locale of the Study

For the purpose of the study, all the four sub-divisions of Palghat district and the three sub-divisions of Alleppey district were included. In the case of Ernakulam district, the rice variety released from RRS Vyttila is meant for pokkali cultivation and pokkali area is confined to Alwaye sub-division. Hence, for the purpose of the study only Alwaye sub-division was selected from Ernakulam district.

At the headquarters of each sub-division there were one Sub-Divisional Agricultural Officer (SDAO) and four SMSs. So all the eight SDAOs and 32 SMSs of the eight sub-divisions viz. Alleppey, Chengannoor, Mavelikara, Alwaye, Shoranur, Mannarghat, Chittur and Alathur were included for the study.

At the AE unit level, the AOs and the ADs were the extension personnel and their number varied in each sub-division depending upon the number of farm families to be covered by each AD. For the purpose of this study, the list of AOs and ADs working in each sub-division was collected from the SDAO's office. There were a total of 156 AOs and 406 ADs in all the eight sub-divisions. To make the sample size wieldy, 50 per cent of AOs and 20 per cent of ADs (78 AOs and 82 ADs) were selected using probability proportionate to size random sampling procedure, from the list. Thus the total number of extension personnel selected from the ESS and the number of extension personnel responded are shown in Table 3.2.

TABLE 3.2: Extension personnel-respondents for the study

Category	No. in position	No. selected	No. responded
SDAOs	8	8	7
SMSs	32	32	28
AOs	156	78	54
ADs	406	82	45
Total	602	200	134

These extension personnel numbering 134 formed the respondents of the study representing the ESS.

3.3.3 *Client Sub-system*

Contact farmers of the eight sub-divisions who had tried at least once and continued to grow the improved rice varieties released by the KAU formed the population of the CSS.

The number of contact farmers in each sub-division varied from 4000 to 4800. So it was considered more convenient to draw samples from the AE units where the number of contact farmers was quite wieldy. The number of AE units in each sub-division also varied from six to eight and there were 55 AE units in the eight sub-divisions having maximum coverge for improved rice varieties. To make the sample manageable, two contact farmers were selected randomly from each AE unit. Thus 110 contact farmers from eight sub-divisions formed the respondents of the study representing the CSS. The number of contact farmers selected from each sub-division is sown in Table 3.3.

TABLE 3.3: Sub-division-wise distribution of contact farmers selected for the study

Name of sub-division	Number of contact farmers selected
Chittur	16
Shoranur	14
Mannarghat	12
Alathur	10
Alwaye	14
Alleppey	16
Chengannur	14
Mavelikara	14
Total	110

3.3.4 *Input sub-system*

Dealers of fertilizer, plant protection chemicals and equipment and representative of credit institutions of Palghat, Ernakulam and Alleppey districts formed the respondents of the ISS. Addresses of 60 personnel of the input supplying agencies were collected and questionnaires were mailed to them along with a self-addressed stamped envelopes. But only 20 of them responded at the first instance. Through reminders and constant personal contacts, 12 more questionnaires could be received back making the total number of respondents from the input sub-system to 32. The total number of respondents from the different sub-systems are shown in Table 3.4.

TABLE 3.4: Number of respondents selected from different categories

Sl. No.	Category	No. of respondents	Percentage to total No. of respondents
1.	Research sub-system	52	15.85
2.	Extension sub-system	134	40.85
3.	Client sub-system	110	33.54
4.	Input sub-system	32	9.76
	Total	328	100.00

3.4 Selection of Variables

3.4.1 *Structural linkages between and among the sub-systems involved in the TOT of improved rice varieties released by the KAU*

The study envisages an analysis of the structural linkages between and among the sub-systems involved in TOT of improved rice varieties released by the KAU. For this purpose, an attempt has been made to collate all available information relating to the structural linkages between and among these sub-systems, in Kerala State. Besides review of literature and official documents, the researcher held discussions with the representatives from each of these four sub-systems and the identified form of structural linkages have been presented and critically analysed elsewhere.

3.4.2 *Functional linkages between and among the sub-systems involved in the TOT of improved rice varieties released by KAU*

Considering the objectives, of the study and based on the review of literature [Jaiswal and Arya (1981); Sen (1984) and Kishore (1986)] it has been conceptualised that research sub-system (RSS), extension sub-system (ESS), client sub-system (CSS) and input sub-system (ISS) constituted the sub-systems in the TOT of improved rice varieties released by the KAU. These four sub-systems cannot function in isolation. For effective TOT, a strong inter sub-system linkage is of vital significance. The possible number of combinations with regard to extent of linkage between and among the four sub-systems will be 12. It was also hypothesised that there would be variation in the extent of linkage activities between these combinations. Therefore, the following 12 possible combinations of linkage among the different sub-systems were considered as the variables for the present study:

1. Extent of linkage of RSS with ESS (RSS-ESS linkage)
2. Extent of linkage of RSS with CSS (RSS-CSS linkage)
3. Extent of linkage of RSS with ISS (RSS-ISS linkage)
4. Extent of linkage of ESS with RSS (ESS-RSS linkage)
5. Extent of linkage of ESS with CSS (ESS-CSS linkage)
6. Extent of linkage of ESS with ISS (ESS-ISS linkage)
7. Extent of linkage of CSS with RSS (CSS-RSS linkage)
8. Extent of linkage of CSS with ESS (CSS-ESS linkage)
9. Extent of linkage of CSS with ISS (CSS-ISS linkage)
10. Extent of linkage of ISS with RSS (ISS-RSS linkage)
11. Extent of linkage of ISS with ESS (ISS-ESS linkage)
12. Extent of linkage of ISS with CSS (ISS-CSS linkage)

The extent of linkage between any two sub-systems is governed by the frequency of performance of the linking activities by the constituents of these sub-systems. In order to exhaust the field of linkage activities in the 12 combinations listed, the following procedure was followed:

Selection of linkage activities under each variable: Based on the review of literature and discussions with experts, a list of all the activities that could possibly establish a linkage between two sub-systems (for the 12 combinations) was prepared. These activities were subjected to relevancy rating among judges adopting the procedure followed by Prasad (1983) on a five-point continuum from most relevant to not at all relevant with scores ranging from 5 to 1. The complete list of activities with proper instructions for judgement was sent to 85 judges comprising extension personnel and scientists working in the various Agricultural Universities and ICAR institutions in India. The judges were asked to express their opinion about the degree of relevancy of each listed linkage activity for inclusion in the study, by marking '√' against each activity in the relevant continuum. Out of the 85 judges only 48 responded. Of these three responses were incomplete and hence rejected. Thus only 45 responses were finally considered for the study. Each linkage activity was compared on the basis of relevancy weightage for deciding its inclusion in the study.

Relevancy weightage was worked out as follows:

$$\text{Relevancy weightage} = \frac{\text{Actual score obtained by an activity}}{\text{Maximum possible score that activity could secure}}$$

Thus the relevancy weightage could range from 0.2 to 1.0. In this study, the weightage secured by the linkage activities ranged from a minimum of 0.63 and a maximum of 0.95. Since it was found difficult to select the linkage activities under each variable based on relevancy weightage which were quite high for all activities, a different criterion was adopted. The average of 0.63 and 0.95 was taken as the cutting point for the selection of linkage activities for inclusion in the study. Thus all those activities which secured a relevancy weight of 0.79 and above were selected for inclusion in the study.

The number of linkage activities originally included under each variable and the number of activities finally selected for study are shown in Table 3.5.

TABLE 3.5: Number of linkage activities originally included and number of activities finally selected

Sl. No.	Name of variable	No. of linkage activities originally included	No. of linkage activities finally selected
1.	RSS-ESS linkage	30	24
2.	RSS-CSS linkage	30	24
3.	RSS-ISS linkage	30	23
4.	ESS-RSS linkage	25	21
5.	ESS-CSS linkage	25	19
6.	ESS-ISS linkage	21	16
7.	CSS-RSS linkage	22	17
8.	CSS-ESS linkage	21	16
9.	CSS-ISS linkage	16	12
10.	ISS-RSS linkage	18	13
11.	ISS-ESS linkage	22	18
12.	ISS-CSS linkage	24	19

The selected linkage activities were incorporated in the questionnaire and interview schedule as the case may be.

3.5 Measurements of Variables

3.5.1 *Measurement of Extent of Linkage Functions*

Since the investigation was mainly aimed at studying the existing functional linkages between and among the different sub-systems in the TOT of improved rice varieties released by the KAU, the respondents of the respective sub-systems were asked to indicate their response to the extent of performance of each linkage activity by them on a five point continuum viz. always, most frequently, frequently, sometimes and never. The scoring procedure used for the purpose was as follows:

Category	Scores
Always	4
Most frequently	3
Frequently	2
Sometimes	1
Never	0

The scores obtained by the respondents under each sub-system for its linkage activities with every other sub-systems were added together and intercorrelations among the four sub-systems were worked out.

3.5.2 *Computation of Mean Linkage Score and Hypothesis Testing*

For testing the hypotheses set for the study, the mean linkage score of the respondents of each sub-system was computed. Then, the mean score of each respondent was ranked and added together. To determine whether the rank totals differ significantly among themselves or not, the Friedman two-way analysis of variance test was applied and based on the results, the hypotheses set for the study were either accepted or rejected.

3.5.3 *Perception about the importance of linkage activities between and among research, extension, client and input sub-systems — sub-system-wise analysis*

To test the hypotheses set for the study based on the degree of importance of linkage functions as perceived by the respondents of each sub-system, they were asked to indicate the perceived importance of linkage functions performed by them on a three-point continuum viz. most important, less important and least important. Scores were given at 3, 2 and 1 respectively for the responses.

For the purpose of the study, a linkage index was developed by adding the scores secured by the respondents under each sub-system and worked out a weighted average score for each respondent using the perception score for different linkage activities included under each sub-system. Comparison among the different sub-systems was made on the basis of the sums of the ranks for perception of importance of linkage activities of each sub-system using the Friedman two-way analysis of variance test and based on the results, the hypotheses set were either accepted or rejected.

3.5.4 *Perception about the importance of linkage activities between and among research, extension, client and input sub-systems — activity-wise analysis*

It was considered desirable to find out the activity-wise perception of importance of the linkage activities between and among the four sub-systems included in the study. For this purpose, the perception scores assigned by the respondents to each of linkage activities included in the four sub-systems were added together and the means were worked out. On the basis of these mean scores the activities were ranked for further interpretation.

3.5.5 *Constraints in rvolving and spreading improved rice varieties as perceived by the research sub-system*

This formed a part of one of the objectives of the study itself.

To identify the constraints, initially the 55 scientists of the four selected major research stations under KAU, where research works on rice are being done, were requested to send few important constraints in evolving and spreading improved rice varieties as perceived by them. To facilitate mailing of the constraints, self-addressed stamped envelopes were also sent along with the request. The constraints thus obtained were pooled together and screened to avoid duplication. Fifteen constraints in evolving and spreading improved rice varieties as perceived by researchers were finally retained. As the second stage, these pooled constraints were again sent to the same group of scientists and they were requested to rate them on a five point continuum ranging from 'most relevant' to 'least relevant'. Scores were given at 5, 4, 3, 2, and 1 for most relevant, relevant, undecided, less relevant and least relevant responses respectively. Forty two scientists returned the pooled constraints after relevancy rating. The relevancy weights secured by each constraints ranged from 0.35 to 0.85. Taking the mid-value as the cutting point, nine constraints which secured relevancy weight of 0.60 and above were finally selected. These nine constraints were incorporated in the questionnaire for research sub-system and the respondents were asked to assign them scores ranging from zero to 100 according to their importance as perceived by them. Based on the scores secured by each constraint they were ranked from 1 to 9 by composite ranking method.

3.5.6 *Constraints in adopting and spreading improved rice varieties released by KAU as perceived by the extension, client and input sub-systems*

This also formed part of one of the objectives, based on review of literature and discussions with some of the extension personnel, scientist, representatives of the input agencies and progressive farmers, 30 constraints in adopting and spreading improved rice varieties by farmers, were listed out. These constraints were sent to 30 extension personnel and 25 personnel of the input supplying agencies (fertilizer and pesticide dealers, secretaries of co-operative societies and Agricultural officers of Nationalised Banks) in the three sub-divisions in Alleppey district for relevancy rating. The respondents were asked to express the relevancy of the statement on a five point continuum ranging from 'most relevant' to 'least relevant'. Self-addressed stamped envelopes were also enclosed for early return of the tool. Similarly, 30 farmers selected at random from the three sub-divisions were contacted in person and their perception about the relevancy of the statements as expressed by them was recorded against each on the appropriate column by the researcher.

Twenty six out of the 30 extension personnel and 20 out of 25 personnel of the input supplying agencies returned the statements after relevancy rating. Thus out of total 85 respondents from the three sub-systems, 76 responded. Scores of 5, 4, 3, 2 and 1 were given for most relevant, relevant, undecided, less relevant, and least relevant responses, respectively as followed by Prasad (1983). Based on judgements of the respondents, relevancy weight was worked out for each statement. The relevancy weightage for the statements varied from 0.48 to 0.88. Taking the mid-value as the cutting point, all those statements which secured a relevancy weight of 0.68 and above were selected for the purpose of this study.

Thus there were 12, 11 and 13 constraints as perceived by extension,client and input sub-systems and respectively which secured a relevancy weight of 0.68 and above. Now, nine constraints common to all the three sub-systems were selected. These were finally incorporated in the interview schedule of the ESS, CSS and ISS. The respondents were asked to assign score to the statements from zero to 100 according to their importance as perceived by them. Thus the constraints were ranked from one to nine based on composite ranking.

3.6 **Procedure Adopted for Data Collection**

Data were collected from the four categories of respondents viz. research, extension, client and input sub-systems. Data from research, extension and input sub-systems were collected with the help of pre-tested structured questionnaires in English. For the client sub-system, a pre-tested interview schedule in Malayalam was developed. The questionnaires and interview schedule consisted of statements for eliciting responses for:

a) the extent of linkage activities performed by each sub-system with every other three sub-systems and
b) the perception of the respondents about the importance of these linkage activities.

Respondents of the research sub-system were contacted in person by the researcher at the respective research stations and got the questionnaires filled up by them.

For collection of data from the extension sub-system, the researcher visited the concerned sub-division headquarters and contacted the SDAOs and SMSs in person. The required number of questionnaires were handed over to the SDAOs for distribution among other officers and the filled in questionnaires were received back through mail. The visits also helped to build a better rapport between the researcher and the extension personnel and also explain in person the method of filling up the questionnaire.

Data from input sub-system were collected through mailed questionnaire. Self-addressed stamped envelopes were also enclosed for returning the filled in questionnaires.

The selected respondents of client sub-system were contacted in person with the assistance of the local AD. The responses elicited were recorded on the schedule.

The data collected were coded, tabulated and fed to a VERSA-IWS computer for analysis.

3.7 Statistical Tools used

3.7.1 *Factor Analysis*

To surmount the difficulty imposed by the unwieldiness of large number of linkage activities under each sub-system, it will be advantageous to identify among them congregations such that the activities in each cluster represent in common, though broad, a particular aspect and are closely interrelated. Factor analysis is a statistical technique which enables to replace a large number of indices which may have very little theoretical meaning with a much smaller number of conceptual variables which make very good sense theoretically (Blalock, 1960).

According to Overall and Klett (1972) factor analysis is a powerful statistical method which aims at explaining the relationships among numerous connected variables in terms of a relatively few underlying variables called factors. This is useful in analysing the dependent structure of a multivariate population when no a priori pattern of causalty is available. The main aim of factor analysis is parsimony of description of a complex structure.

Consider a multivariate system consisting of '*p*' responses described by the random variables x_1, x_2 .... x_p. The x_i are supposed to have a multivariate normal distribution. Then the basic model of factor analysis is given by

$$X_j = \Sigma_r^m = il_{jryr} + l_j$$

$$(j = 1, 2, \ldots p)$$

where *yr* = r^{th} common factor variate (r^{th} factor)

ljr = parameter representing the importance of the rth factor in combination with the jth response known as the factor loadings of the jth response on the rth common factor

lj = j^{th} specific factor variate

Let the common factor variates be independently and normally distributed with zero means that unit variance. Then it is evident that

$$S_{j}^{2} = \Sigma_{r}^{m} = i l_{jr}^{2} + U_{j}^{2}$$

where Sj^2 = Total variance of the jth response variates

Uj^2 = Specific variance of the jth response variates

$\Sigma_{r}^{m} = 1^{\mu} ljr = h^2 =$ communality

Communality of a variable is the sum of its independent common variances i.e. a measure of variance of each variable with all other variables and is represented by the symbol h^2 (Fruchter, 1954).

The parameter *ljr* can be shown to be as the correlation between j^{th} response and the rth common factor. Thus factor loadings can be regarded as the correlation coefficient between the original variables and the various common factor variables or factors.

The most important aspect of factor analysis is the extraction of factor loadings. Several methods can be adopted for the extraction of factor loadings. Among them, the centroid method, the principal axis or principal-factor analysis method and the maximum likelihood method are the more frequently employed techniques. In the present study, the principal-factor analysis method as described by Harman (1960) was used because of its elegancy and similarity to the principal component analysis.

The mechanics of extracting factors involve initially filling in the principal diagonal of the correlation matrix with suitable values of communalities. Thurstone (1947) suggested a simple method which consisted in choosing the highest correlation which the variable in question would have with any of the other variables with the set, as the communality value. The principal-factor analysis may be repeated using the set of communalities obtained from the first solution. This procedure of iteration can be repeated as often as necessary, until the communalities arising from the successive analysis no longer differ significantly. This procedure has been followed in this study.

As the preliminary step to more elaborate calculation, the zero order correlation matrix (r) of the original linkage activities was worked out. The characteristic roots or eigen values obtained from the determinant equation $|R\text{-}\lambda I| = 0$, where, I is the identity matrix and 0 is the null matrix. 'R' will have a maximum of 'P' eigen values which may be non-negative. If $\lambda_1, \lambda_2 \ldots \lambda_p$ are the characteristic roots, then $\Sigma^P\lambda_i$ = Trace $R = P$. The eigen values smaller than one may be neglected (Geer, 1971).

The eigen vectors (ai) corresponding to the eigen value λ_i is defined by the equation $(R-\lambda_i I)\ ai = 0$. The characteristic roots measure the amount of variability explained by the different factors. The factor loading of the jth response on the rth factor is given by $l_{jr} = a_{jr}\sqrt{\lambda r}$ proportionate variance contributed by the jth factor is given by $V = \lambda r/p$

If a variable has no significant correlation with a component, then that variable is not contributing significantly to the variance of the component. If the variable is not correlated with any of other components, then it can be eliminated as contributing nothing towards total variability. Fruchter (1954) suggested a limit of 0.5 and variables having factor loadings from 0.5 to 0.7 were considered as high and above 0.7 as very high. Walton (1972) found a minimum correlation co-efficient (factor loadings) of 0.45 among traits affected by a common factor. Asawa (1981) in his studies on factor analysis in chick pea restricted to select only those characters which gave a factor loading greater than or equal to 0.50 to provide for meaningful representation of the relationship among original variables. In this study 0.45 has been taken as the minimum limit and hence linkage activities having factor loading 0.45 and above only were considered for interpretation.

The variables which do or do not relate substantially for eacn factor should be clearly distinguishable. For this, the factor axes have to be redefined to make the resulting factors mutually orthogonal. Such a transformation is expected to make the loadings more meaningful and interpretable. Kaiser's (1958) raw varimax criterion for simple structure has been widely used for rotation. In this study also, the eigen vectors and factor loadings were worked out by using an iterative alogarithm for the principal-axes method using a VERSA IWS computer. Kaiser's varimax rotation was also incorporated in the programme.

3.7.2 *Development of a Linkage Index*

A weighted average score for each respondent was worked out by computing the weighted mean of linkage scores, the weight being the mean perception score for importance of different linkage activities included under each sub-system. The formula used for computation is given below:

$$1_{wkj} = \frac{\sum_{i=1}^{mk} wik\ lijk}{\sum_{i=1}^{mk} wik}$$

where *lwkj* = weighted linkage score of the jth respondent in the kth sub-system

mk = no. of linkage activities in the kth sub-system (k = 1, 2 or 3)

lijk = the linkage score of the jth respondent for the ith linkage activity in the kth sub-system

wik = $\dfrac{\sum_{i=1}^{nk} pik}{nk}$

where *pik* = perception score for the ith linkage functions of the kth sub-system

nk = no. of respondents in the kth sub-system

In order to give allowance for the varying number of linkage activities under each sub-system, percentage of effective linkage score was worked out by using the formula,

$$PcL = \frac{Lwkj}{mk} \times 100$$

Linkage activities of the respondents in different sub-systems were then ranked based on linkage index score and the ranks of the respondents with regard to the different sub-systems were arranged in a two-way table.

3.7.3 *Friedman Two-way Analysis of Variance by Ranks*

The Friedman test was applied to determine whether the rank total of the different sub-systems (*Rj*) differ significantly among themselves or not. To make this test, the value of a statistic which Friedman (1937) denoted as $\chi^2 r$ was computed. When the number of rows and/or columns was not too small, it can be shown that $\chi^2 r$ is distributed approximately as chi-square with $df = k-1$.

$$\chi_r^2 = \left[\frac{12}{Nk(k+1)} \sum_{j=1}^{k} (Rj)^2\right] - 3N(k+1)$$

Where, N = Number of rows
k = Number of columns
Rj = sum of ranks in j^{th} column directs to sum the squares of the sum of ranks over all k conditions.

3.7.4 *Critical Value for Making Multiple Comparisons*

A significant $\chi^2 r$ value indicates that there was significant difference among different sub-systems with regard to their linkage activities. In such a situation, the next step will be to make a comparison among the different sub-systems with regard to the linkage efficiency. Zar (1981) has given a method of applying the Student Newman (1939) and Keuls (1952) test for comparing rank totals in two-way analysis of variance using ranked data. The method consists in calculating the critical value wi, as given below:

$$wi = Q(nf)\, SE_m$$

where, wi = critical value
$Q(nf)$ = table value of studentized range statistic for 'n' systems and '*f*' degrees of freedom of error
SEm = standard error of mean

All possible differences among rank totals of different sub-systems will have to be enumerated and compared with the relevant critical values. A range of difference in '*i*' sub-system rank totals in the array will be compared by using wi $(c = 2 \ldots k)$ for assessing significance.

3.8 **Operationalisation of Concepts**

According to Kerlinger (1965) operational definition is a specification of the activities of the researchers in measuring a variable or in manipulating it. He further suggested that an operational definition is a sort of manual of instructions to the investigator.

Rogers and Shoemaker (1971) stated that it might be a scale, index, observation or the answer to a direct question.

Based on the review of literature, for the purpose of this study, the concepts were operationally defined as follows:

3.8.1 *Linkage*

Linkage is operationalised as the working relationship between any two sub-systems.

3.8.2 *Structural Linkage*

Refers to the formal institutional arrangements/mechanism constituted for the purpose of establishing effective liaison between and among the four sub-systems involved in the transfer of technology of improved rice varieties.

3.8.3 *Functional Linkage*

Functional linkage is operationally defined as the extent of working relationship between and among the four sub-systems in the transfer of technology of improved rice varieties.

3.8.4 *Research Sub-system*

The farm scientists of the Kerala Agricultural University who are involved in evolving the developing improved rice varieties constitute the research sub-system.

3.8.5 *Extension Sub-system*

The officers of the department of agriculture from the sub-divisional level to the village level who are responsible for transfer of technology of improved rice varieties constitute the extension sub-system.

3.8.6 *Client Sub-system*

The Contact farmers of the study area (under the T&V system) who had at least tried once and those who continue to grow the improved rice varieties released by KAU constitute the client sub-system.

3.8.7 *Input Sub-system*

The agencies who are responsible for the distribution of fertilisers, pesticides and credit to the farmers of the study area form the input sub-system for the study.

3.8.8 *Transfer of Technology*

Transfer of technology is operationally defined as the spreading of recommended cultivation practices of the improved rice varieties released by the KAU to the farmers through the different sub-systems.

3.8.9 *Improved Rice Varieties*

For the purpose of this study, improved rice varieties are operationally defined as the rice varieties evolved and released from the RARS Pattambi, RRS Moncompu, Vyttila and Kayamkulam after the inception of the KAU.

3.8.10 *Systems Analysis*

It is operationalised as the systematic way of identifying the method of interacting with each other and the different steps taken to move towards this interaction by the research, extension, client and input sub-systems taken each sub-system as a whole in the transfer of technology of improved rice varieties.

3.9 General Hypotheses set for the Study

The following general hypotheses were formulated for the study and these were tested using the procedure described as 4.2 and 4.3 in chapter 4 under measurement of variables.

3.9.1 *Extent of Linkage Activities Performed by Respondents*

3.9.1.1 The extent of linkage of research sub-system with extension sub-system would be on par with that of client and input sub-systems.

3.9.1.2 The extent of linkage of research sub-system with client sub-system would be on par with that of extension and input sub-systems.

3.9.1.3 The extent of linkage of research sub-system with input sub-system would be on par with that of extension and client sub-systems.

3.9.1.4 The extent of linkage of extension sub-system with research sub-system would be on par with that of client and input sub-systems.

3.9.1.5 The extent of linkage or extension sub-system with client sub-system would be on par with that of research and input sub-systems.

3.9.1.6 The extent of linkage of extension sub-system with input sub-system would be on par with that of research and client sub-systems.

3.9.1.7 The extent of linkage of client sub-system with research sub-system would be on par with that of extension and input sub-systems.

3.9.1.8 The extent of linkage of client sub-system with extension sub-system would be on par with that of research and input sub-systems.

3.9.1.9 The extent of linkage of client sub-system with input sub-system would be on par with that of research and extension sub-systems.

3.9.1.10 The extent of linkage of input sub-system with research sub-system would be on par with that of extension and client sub-systems.

3.9.1.11 The extent of linkage of input sub-system with extension sub-system would be on par with that of research and client sub-systems.

3.9.1.12 The extent of linkage of input sub-system with client sub-system would be on par with that of research and extension sub-systems.

3.9.2 *Perception of Respondents About the Importance of Linkage Activities*

3.9.2.1 The perception about the importance of linkage activities of research sub-system with extension sub-system would be on par with that of client and input sub-systems.

3.9.2.2 The perception about the importance of linkage activities of research sub-system with client sub-system would be on par with that of extension and input sub-systems.

3.9.2.3 The perception about the importance of linkage activities of research sub-system with input sub-system would be on par with that of extension and client sub-systems.

3.9.2.4 The perception of importance of linkage activities of extension sub-system with research sub-system would be on par with that of client and input sub-systems.

3.9.2.5 The perception of importance of linkage activities of extension sub-system with client sub-system would be on par with that of research and input sub-systems.

3.9.2.6 The perception of importance of linkage activities of extension sub-system with input sub-system would be on par with that of research and client sub-systems.

3.9.2.7 The perception of importance of linkage activities of client sub-system with research sub-system would be on par with that of extension and input sub-systems.

3.9.2.8 The perception of importance of linkage activities of client sub-system with extension sub-system would be on par with that of research and input sub-systems.

3.9.2.9 The perception of importance of linkage activities of client sub-system with input sub-system would be on par with that of research and extension sub-systems.

3.9.2.10 The perception of importance of linkage activities of input sub-system with research sub-system would be on par with that of extension and client sub-systems.

3.9.2.11 The perception of importance of linkage activities of input sub-system with extension sub-system would be on par with that of research and client sub systems.

3.9.2.12 The perception of importance of linkage activities of input sub-system with client sub-system would be on par with that of research and extension sub-systems.

4

RESULTS AND DISCUSSION

THE RESULTS of the study are presented and discussed in this chapter under the following heads:

4.1 Structural linkages between and among the research, extension, client and input sub-systems in the TOT of improved rice varieties released by the KAU.

4.2 Extent of functional linkage between and among the different sub-systems in the TOT of improved rice varieties released by the KAU — results of Friedman two-way analysis of variance by ranks:

- 4.2.1 RSS with ESS, CSS and ISS
- 4.2.2 ESS with RSS, CSS and ISS
- 4.2.3 CSS with RSS, ESS and ISS
- 4.2.4 ISS with RSS, ESS and CSS

4.3 Factors associated with the TOT of improved rice varieties released by KAU in respect of research, extension, client and input sub-systems — results of principal-factor analysis.

- 4.3.1 RSS with ESS
- 4.3.2 RSS with CSS
- 4.3.3 RSS with ISS
- 4.3.4 ESS with RSS
- 4.3.5 ESS with CSS
- 4.3.6 ESS with ISS
- 4.3.7 CSS with RSS
- 4.3.8 CSS with ESS
- 4.3.9 CSS with ISS
- 4.3.10 ISS with RSS

4.3.11 ISS with ESS
4.3.12 ISS with CSS

4.4 Importance of linkage activities between and among research, extension, client and input sub-systems as perceived by the respondents — sub-system-wise analysis.
4.5 Activity-wise analysis.
4.6 Constraints in evolving, spreading and adopting improved rice varieties released by the KAU as perceived by different sub-systems.

4.6.1 Constraints in evolving and spreading improved rice varieties released by the KAU as perceived by RSS.
4.6.2 Constraints in spreading and adopting improved rice varieties released by the KAU as perceived by the extension, client and input sub-systems.

4.7 Empirical model of the linkages between and among the different sub-systems in the TOT of improved rice varieties released by the KAU.
4.8 Suggested model of linkages for transfer of agricultural technology.

4.1 Structural linkages between and among the research, extension, client and input sub-systems in the TOT of improved rice varieties released by the KAU

In this study, structural linkages refer to the formal and established mechanisms and/or forums which facilitate coordinated activity among the four sub-systems identified for the TOT of improved rice varieties released by the KAU. With specific reference to rice, the RSS encompasses exclusively the KAU which has the prime responsibility in this respect in Kerala. The ESS is largely confined to the State Department of Agriculture, Kerala which shoulders the task of spreading improved rice varieties among the rice growers of the State. The ISS is represented by the multiplicity of Nationalised Banks, Service Co-operative Banks/Societies and Fertilizer-Pesticide-Weedicide firms active in the State. The CSS exhausts the rice growing farmers of the State. An effort was made to identify and analyse the mechanisms and/or forums which foster co-ordinated activities between and among these four sub-systems in the State for the TOT of improved rice varieties released by the KAU and the information is furnished hereunder.

4.1.1 *RSS*

As KAU is the only institution representing the RSS involved in the development of improved rice varieties in the State of Kerala, the structural linkages existing in the KAU are identified and analysed in the pages that follow:

The KAU

The KAU came into existence in 1972 with the enactment of the KAU Act 33/1971. It has a State-wide mandate for the triple functions of agricultural research, teaching and extension education and it assumed the role of the principal agency to provide the leadership in agricultural research activities in the State.

The KAU is a multicampus University. All the research work under the University is being carried out by the scientists and post-graduate students in the five teaching campuses at Vellanikkara, Mannuthy, Vellayani, Panangad and Tavanur as well as in the 23 research stations under the University located in different parts of the State.

Research Organisation

The KAU has initiated a research policy which emphasises the need for research, oriented towards problems of farmers and to substantially improve their standard of living, thereby ultimately aiming at the increase in the total agricultural production of the State as well as the net income of the farmers.

In consonance with the above research policy, the research organisation is structurally linked with teaching and extension education. The organisational set up of research administration of the KAU is schematically presented in Fig. 4.1.

The research projects of the University consist of State Government sponsored research schemes (Departmental projects), All-India Coordinated Research Projects and *ad hoc* research projects of ICAR and other externally aided research projects.

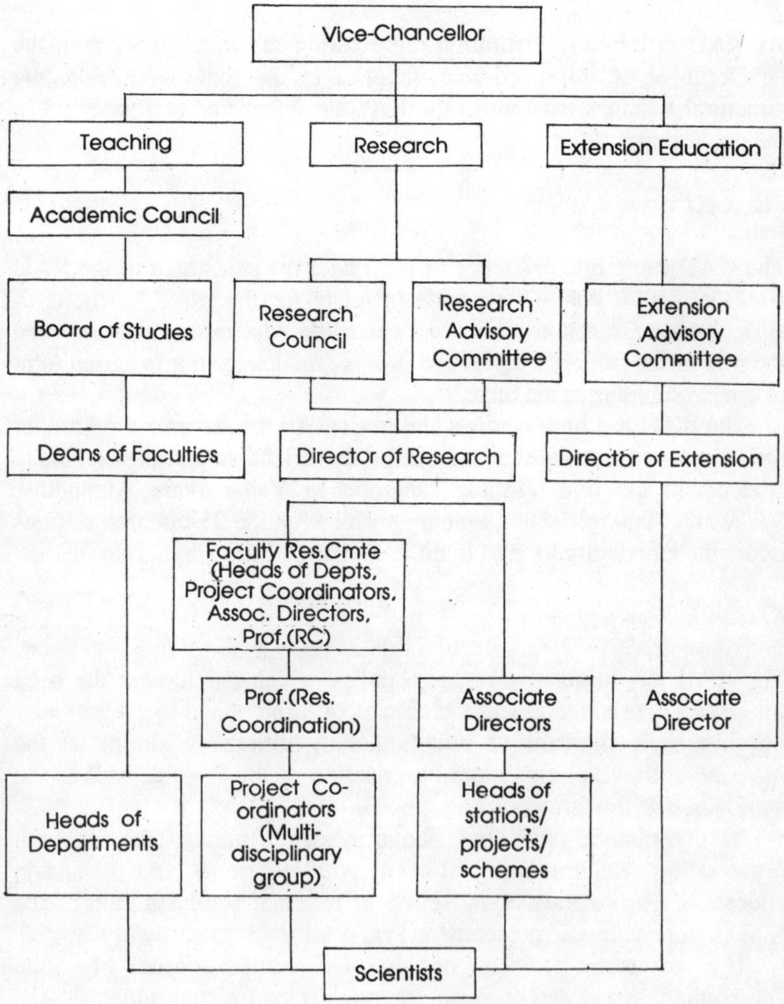

FIG. 4.1: Diagrammatic representation of the research administration in KAU

Formulation and Approval of Research Programmes

The research projects prepared by the scientists are passed on to the Project Coordinators through the heads of stations/projects. The Project Coordinator scrutinises the research projects with the help of multi-disciplinary groups of scientists and passes them to the Faculty Research Committee (FRC). The FRC meets as and when required and makes a detailed analysis of the research projects for technical perfection, feasibility, viability, resource utilization, etc. After approval, the research projects are routed to the Director of Research who issues administrative sanction to the projects for implementation.

The Director of Research also places the research projects before the Research Council which formally approves the projects.

The Research Council consists of the Development Commissioner and the Heads of other major development departments like Agriculture, Animal Husbandry, Dairy Development, Fisheries, Forestry and representatives of ICAR institutions in addition to selected Scientists of the KAU. The schematic representation of the process of formulation and approval of the research programmes of the KAU is given in Fig. 4.2. Suggestions from the T&V monthly workshops, NARP and T&V zonal workshops, farmers' group discussions etc. are also taken into consideration while formulating the research projects.

Thus structural linkage of the RSS with other sub-systems starts even at the stage of formulation of research projects.

Extension Education Organisation

The extension education activities of the KAU are coordinated by the Director of Extension. The extension education policy of the University is based on the following principles.

a) There is an adequate net-work of field extension functionaries in the State departments and University need not duplicate their services.
b) The literacy and wide newspaper reading habit of the farming community must be taken into consideration while organising extension education programmes.
c) It will be desirable to provide the educated farmers not only technical information and skills in farming but also make them practising scientists by imparting information which form the basis of modern technology.

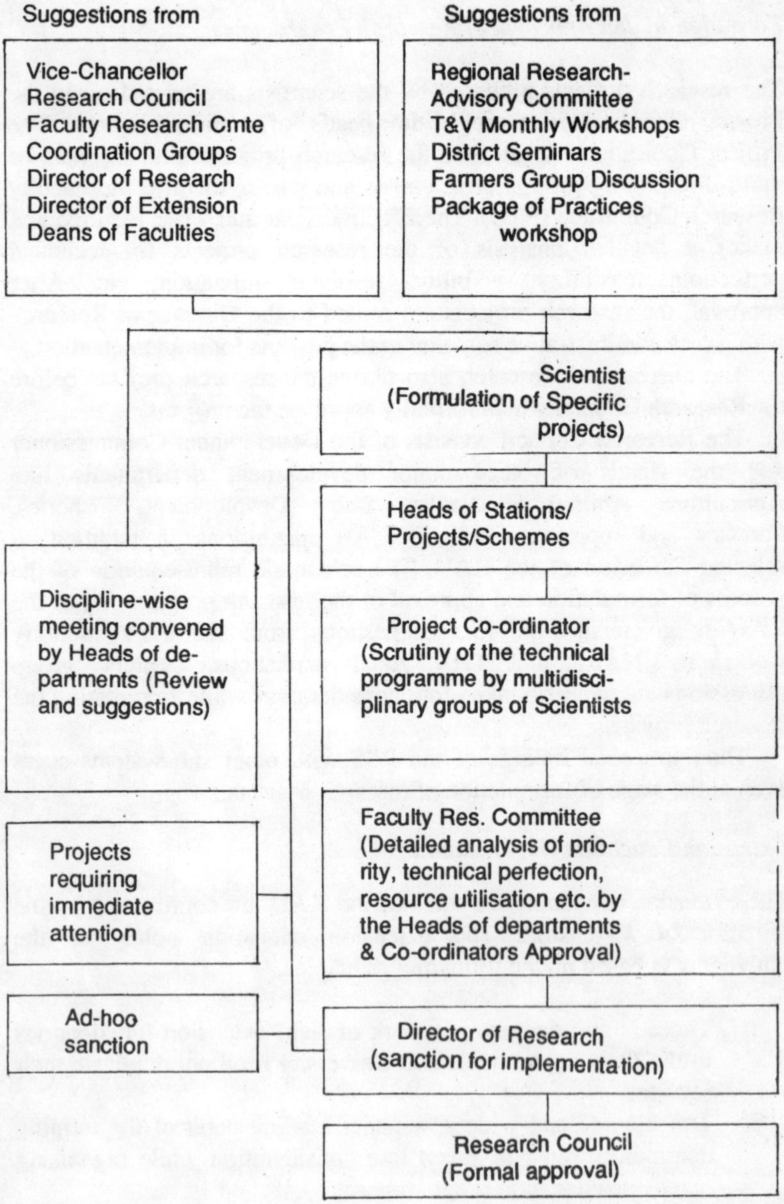

FIG. 4.2: Diagrammatic representation of the process of formulation and approval of research projects in KAU

i) Extension Advisory Committee (EAC)

In the formulation and implementation of extension education programmes of the KAU, the Director of Extension is being advised by the EAC. In order to orient the extension activities based on the needs and aspirations of the farming community, the EAC is constituted giving representation to various development departments of the State like Agriculture, Animal Husbandry, Fisheries, Rural Development, Forests, Cooperation and persons of eminence and progressive farmers. The Director of Research and the Deans of different faculties are also members of the EAC. The Committee is headed by the Vice-Chancellor and meets annually to review the activities of the previous year and formulate the extension education programmes for the next year.

ii) Joint Training Committee (JTC)

The JTC is the apex body constituted by the KAU to assess the training needs of the extension personnel working in various development departments of Kerala. Thee Vice-Chancellor is the Chairman of the JTC and the Director of Extension is the Member-convenor. The Director of Research, the Deans of Faculties and the heads of State Development departments like Agriculture, Animal Husbandry, Rural Development, Forests, Cooperation, Dairy Development, Education and fisheries are members of the Committee. The Committee meets annually to finalise the training courses to be organised by the KAU to cater to the needs of the development departments.

iii) Extension Education Activities of the KAU

Since the KAU has the responsibility of developing and trying innovative extension approaches, a number of such programmes are being implemented at present. Through many of these approaches the KAU is also involved in the TOT of improved rice varieties released by it. An attempt is made here to epitomise the structural linkages existing in these programmes.

a) Training programmes
b) Information support
c) Farm advisory service

d) Field extension studies
e) Trials, testing and demonstrations
f) National Demonstration Scheme
g) Village adoption programme
h) Lab-to-land programme

a) Training Programmes

The KAU organises short and medium duration training courses in Agriculture and allied subjects, both for officials and farmers. The Central Training Institute at the Mannuthy campus and the Training Service Scheme at the Vellayani campus take care of the training programmes for officials. Officers of the various development departments of the State are given training in the latest agricultural technology and allied subjects under the auspices of these two training units. Some of these trainings are focused at transferring the technology of improved rice varieties released by the KAU.

The Krishi Vigyana Kendras (KVKs) at Pattambi, Ambalavayal and Manjeswar offer short duration, need-based training to practising and prospective farmers.

The ADs working at the village level under the Department of Agriculture are also given pre-service agricultural training for six months at the College of Agriculture, Vellayani and the RARS Pilicode.

b) Information Support

The latest agricultural technology and the results of research are diffused to the target clientele through the agricultural columns of five leading vernacular newspapers of Kerala, the All India Radio, The Doordarshan Kendras and the KAU publications like *KAU Newsletter*, *Agres News* and the semi-scientific Malayalam publication '*Kalpadhenu*'. Technology transfer is also effected through correspondence course in different subject matter areas.

Participation in major exhibitions, organising *kisan melas*, workshops, seminars, farmers' days and mini exhibitions are the other supportive activities of the KAU for TOT.

c) Farm Advisory Service (FAS)

The FAS attached to the Directorate of Extension Education organises district level seminars on Agriculture and allied subjects in collaboration with development departments, input agencies and voluntary organisations. The FAS consisting of a team of multi-disciplinary Scientists offers advisory services to those who are interested and seek the services of the KAU.

d) Field Extension Studies

The Directorate of Extension has taken up the responsibility of ascertaining the performance of the technologies released by the KAU in the field. Such field extension studies also aim at assessing the impact of these technologies on the farming communities, so as to serve as a feedback mechanism for identifying further research areas.

e) Trials, Testing and Demonstrations

As a prelude to the release of new technology, farm trials and demonstrations of the newly evolved crop varieties and technologies are carried out jointly by the KAU and the Department of Agriculture at Multi-locations in the farmers' fields. These trials also help to get effective feedback from the farmers and to foster TOT process.

f) National Demonstration Scheme (NDS)

One of the major objectives of the NDS is to maximise the per unit area production by applying the latest agricultural technology on the selected crop. The scheme operated with the financial assistance of the ICAR is operative in the selected districts continuously for five years. The demonstrations carried out in the farmers' fields by a team of multi-disciplinary Scientists of the KAU act as a strong link in the TOT of food crops and oil seeds. The Advisory Committee constituted for this programme provides for the involvement of Agricultural Department personnel, the progressive farmers and the Scientists of the University.

g) Village Adoption Programme (VAP)

The KAU has adopted 32 villages adjacent to its five teaching campuses and major research stations. These villages serve as field laboratories for testing the newly developed technologies and for identifying the field constraints, if any. The research results tested in the adopted villages also serve as a social audit of the technology.

h) Lab-to-land Programme (LLP)

The lab-to-land programme implemented as a part of the Golden Jubilee Celebrations of the ICAR has made much impact in the TOT process. The programme is implemented through selected research stations under the KAU, designated as the 'Transfer of Technology Centres'. Marginal farmers and scheduled caste/scheduled tribe farmers are selected on a cluster basis, as the beneficiaries of the programme. They are provided with production inputs like improved seeds, planting materials, improved breeds of poultry, goat, fertilizers, etc. besides capital inputs. To supplement the TOT process, kisan melas are also organised in the villages. The programme is implemented through the coordinated efforts of the research, extension, client and input sub-systems. At present 300 farming families in the TOT centres are benefited by these programmes in the State. Inter-disciplinary teams have been constituted on a regional basis to ensure better coordination among the various development and research personnel.

4.1.2 *Linkage of RSS with ESS*

Any research programme to be field oriented, should be closely linked with the feedback system. This aspect has been given particular emphasis in the zonal concept accepted after the implementation of the NARP. As a corollary to this, all the research efforts of the KAU focus mainly on location-specific problems identified in the monthly and zonal workshops, during the joint field visits of researchers and extension workers, in the farmers' group discussions, seminars etc.

4.1.3 *Regional Research Advisory Committee*

The State has been divided into five agro-climatic zones viz. Northern, Central, Southern, High ranges and the Region of problem areas. Research Advisory Committee has been constituted for each zone with members from among the senior research scientists of the KAU, representatives of the various development departments of the State like Agriculture, Rural Development, Cooperation and Minor irrigation. Representatives of local bodies and progressive farmers are also co-opted as members of the Committee. The Committee holds half-yearly meetings to review the progress in research and to identify the production constraints.

4.1.4 *Monthly Workshops*

The pre-monthly workshop meetings and the monthly workshops under the T&V system, conducted in each district serve as important linking forums of researchers and extension personnel. In these workshops, selected scientists of the KAU function as resource persons. The PAO of the district, the SDAOs and the SMSs of the sub-divisions in the district represent the ESS. In these workshops the crop-weather-input situation is reviewed, the farm trial results discussed, the field problems presented and discussed and production recommendations formulated for the ensuing fortnight. The joint field visits, after the workshop, by the diagnostic team consisting of scientists and extension personnel further fortifies the research-extension linkages.

4.1.5 *Regional Workshops*

The combined zonal workshop (of the NARP and the T&V system) organised once every six months, brings together Scientists of the KAU, the ICAR, the CSIR and other institutions engaged in agricultural research in the region and the extension personnel of the Department of Agriculture of the respective zone. The Workshops review the progress of the research projects in the preceding season with a view to identifying recommendations that can be made for the regional farming situations. In these workshops new research projects are formulated for the succeeding season based on the field requirements and farm trial programmes chalked out.

4.1.6 *State Level Package of Practices Workshop*

The main purpose of this workshop is to modify the Package of Practices Recommendations for the major cultivated crops of the State, based on the latest research findings of the KAU and the ICAR institutions of the State. Scientists of the KAU and the ICAR institutions, extension personnel of State Department of Agriculture and representatives of input supplying agencies are thus brought together in the Workshop. The research results are placed in such workshops for open discussion among the participants. The research results thus discussed and finalised, come out in the form of a compendium of "Package of Practices Recommendations" for crops. Though every year such workshops are envisaged the periodicity is seldom adhered to. Depending upon the availability of research results, the workshops may be annual, biennial and so on.

4.1.7. *Linkage of RSS with CSS*

In addition to the KVKs, and FAS and the NDS, the farm trial programmes, the joint field visits of diagnostic teams, the village adoption programme, the lab-to-land programme, Rice Day celebrations, *Kisan melas*, Seminars, Farmers' days etc. conducted at the research stations are some of the important activities which structurally and functionally link the RSS and the CSS.

4.1.8 *Linkage of RSS with ISS*

Among the input sub-systems, the fertilizer and pesticide dealers have better liaison with the RSS whereas the linkage with credit institutions is very limited. Testing the newer products of fertilizer and pesticide dealers at the research stations, execution of certain sponsored research schemes by scientists etc. are the important means of linkage between the RSS and the ISS. Participation of the ISS in Package of Practices Workshops, seminars and exhibitions organised by the University helps to strengthen, both structurally and functionally, the relation between the RSS and the ISS.

4.1.9 *The Department of Agriculture*

The Department of Agriculture, with its field staff working from village level to sub-division level is the main extension agency in the State involved directly in the TOT of improved rice varieties released by the KAU. Though the Department of Agriculture has undergone some organisational changes recently, the organisation of the department, when the study was taken up, was based on the guidelines given under the T&V system (Fig. 4.3).

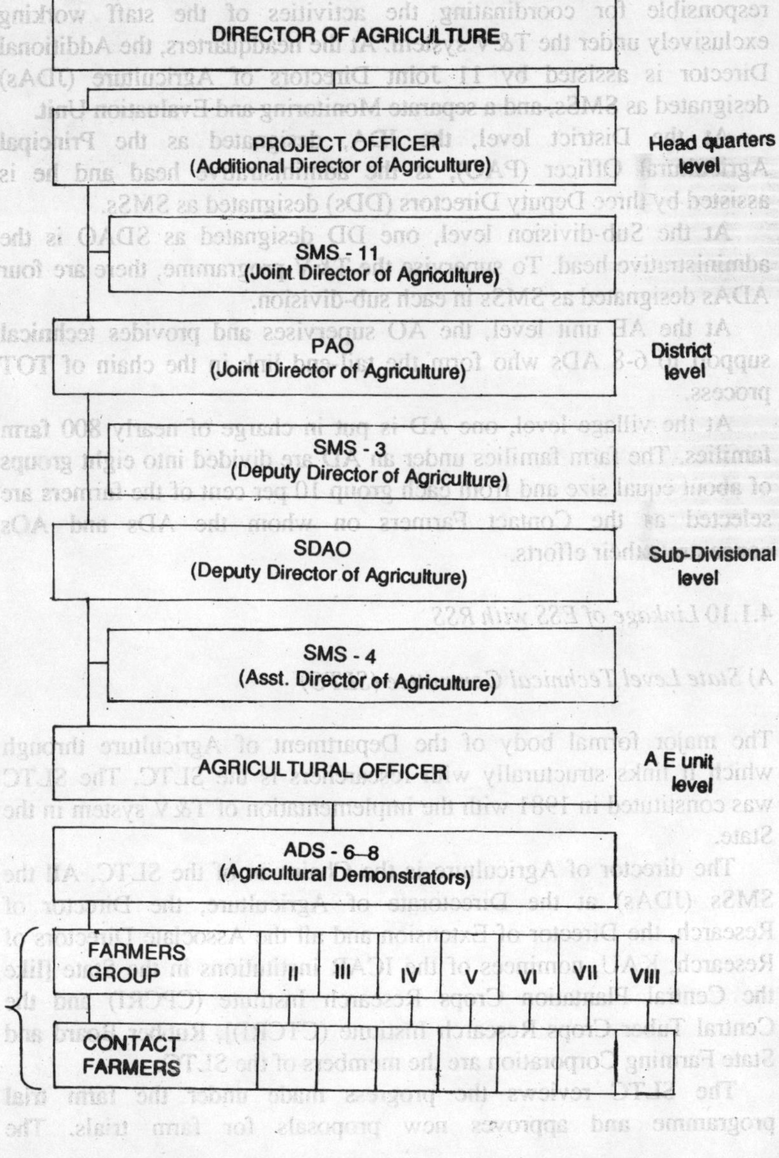

FIG. 4.3: Organisational set-up of the Department of Agriculture (T&V system)

The Director of Agriculture is the administrative head of the Department of Agriculture. At the State level, he is assisted by an Additional Director of Agriculture designated as Project Officer. He is responsible for coordinating the activities of the staff working exclusively under the T&V system. At the headquarters, the Additional Director is assisted by 11 Joint Directors of Agriculture (JDAs) designated as SMSs, and a separate Monitoring and Evaluation Unit.

At the District level, the JDA, designated as the Principal Agricultural Officer (PAO), is the administrative head and he is assisted by three Deputy Directors (DDs) designated as SMSs.

At the Sub-division level, one DD designated as SDAO is the administrative head. To supervise the T&V programme, there are four ADAs designated as SMSs in each sub-division.

At the AE unit level, the AO supervises and provides technical support to 6-8 ADs who form the tail-end link in the chain of TOT process.

At the village level, one AD is put in charge of nearly 800 farm families. The farm families under an AD are divided into eight groups of about equal size and from each group 10 per cent of the farmers are selected as the Contact Farmers on whom the ADs and AOs concentrate their efforts.

4.1.10 *Linkage of ESS with RSS*

A) State Level Technical Committee (SLTC)

The major formal body of the Department of Agriculture through which it links structurally with researchers is the SLTC. The SLTC was constituted in 1981 with the implementation of T&V system in the State.

The director of Agriculture is the Chairman of the SLTC. All the SMSs (JDAs) at the Directorate of Agriculture, the Director of Research, the Director of Extension and all the Associate Directors of Research, KAU, nominees of the ICAR institutions in the State [like the Central Plantation Crops Research Institute (CPCRI) and the Central Tuber Crops Research Institute (CTCRI)], Rubber Board and State Farming Corporation are the members of the SLTC.

The SLTC reviews the progress made under the farm trial programme and approves new proposals for farm trials. The

Committee will also critically analyse the reports of the monitoring and Evaluation cell of the Kerala Agricultural Extension Project (KAEP) and suggest mid-course corrections in the implementation of the KAEP. The SLTC met seven times from 1983 to 1986.

B) District Level Technical Committee (DLTC)

This Committee was also constituted along with the SLTC to ensure structural linkage between the extension personnel, the researchers and farmers at the district level.

The PAO is the Chairman of the DLTC. All SMSs of the district and representatives of farmers are the members of the DLTC.

The DLTC has the mandate to review the recommendations of agricultural practice, to identify research priorities and adaptive research needs for transmission to the SLTC. The Committee is scheduled to convene twice an year.

Since 1985, the DLTC has been dispensed with so as to avoid duplication of efforts as the functions assigned to the DLTC are performed by Zonal Workshops organised under NARP.

4.1.11 *Linkage between ESS and CSS*

The basic extension technique adopted under the T&V system is a systematic programme of training of the AOs and ADs by the SMSs and scheduled visits to the Contact farmers and other farmers by the ADs in order to transfer the latest agricultural technology on the cultivated crops. These visits by the ADs will be supervised by the AOs of the AE unit and the SMSs of the sub-division. The systematic and scheduled visits to the farm families by the extension personnel form the primary and most important structural linkage between ESS and CSS. The seminars, melas, exhibitions, campaigns etc. organised by the ESS also serve as structural forums where the CSS interact with the ESS.

4.1.12 *Linkage between ESS and ISS*

The Fertilizer Standing Committee (FSC) consisting of representatives of the Department of Agriculture and the fertilizer firms, is the only formal body linking organisationally the ESS and ISS. The input dealers' training to be organised by the SDAO is not being done regularly. Participation of the representatives of the input agencies in seminars, *melas*, workshops, exhibitions, etc. organised by the ESS informally links the two sub-systems.

The FSC explores the ways and means of promotion of fertilizer consumption and streamlining the supply system. Pesticides and credit being need-based inputs, the agencies do not have formal link with the ESS. Cash and pesticides being important inputs in crop production, the links between ESS and ISS need to be carefully defined and developed.

4.1.13 *CSS with Other Sub-systems*

The linkages between the CSS and the other sub-systems are need-based rather than structurally organised. It is a fact that there is no client organisation exclusively for rice. But a number of organisations oriented towards other crops/or specific purposes exist in Kerala as elsewhere in the country since long. The *Bharath Karshak Samaj*, the *Kerala Karshaka Sangam*, the Rubber Growers' Association, the *Kole Karshaka Sangam* (Trichur district), *Karshaka Sangams* backed by different political parties etc. are a few of them. These organisations, at times, convene special functions in which the researchers, extension personnel and input agencies are invited to lead technical discussions. Such forums become a venue for the CSS to link itself with the other sub-systems. Thus it could reasonably be concluded that the structural linkage of the CSS with the other sub-systems is sporadic in nature.

4.1.14 *Structural Linkages in the TOT of Improved Rice Varieties: A Bird's Eye View*

The foregoing analysis of the existing structural linkages in the TOT of improved rice varieties/agricultural practices brings into lime light some thought provoking facts.

The KAU, being vested with the prime responsibility of generating improved rice varieties/agricultural practices suited to varied agro-climatic conditions of the State, maintains effective structural linkages with the other research organisations of the State like the CPCRI, the CTCRI, the CSIR, the Cardamom Research Institute of the Cardamom Board and the Rubber Research Institute of India. Likewise, an equally effective linkage is being maintained with the major development departments of the State like Agriculture, Animal Husbanary, Rural Development, Forest, Fisheries, cooperation, etc. Among them, the linkage between the KAU and Department of Agriculture deserves special mention since it is the primary agency responsible for the TOT of improved practices of rice to the target clientele through a network of field extension personnel. The clientele, in turn, are directly and indirectly linked with the KAU research sub-system and the extension organisation, particularly the Department of Agriculture. These linkages have been ensured in the KAU by giving due representation to other research organisations, development departments and progressive farmers in its statutory bodies like the General Council and the Academic Council and non-statutory bodies like the Research Council, the Zonal Research Advisory Committee and the EAC. This states in unequivocal terms that the KAU maintains the most effective structural linkage with these agencies. With the implementation of the NARP by the KAU and the T&V system of extension work by the Department of Agriculture, it is heartening to note that the linkage between these two sub-systems has strengthened many-fold facilitating symbiotic relationship. The Monthly Workshops, the Zonal Workshops, meetings of the SLTC, the State Level Package of Practices Workshops etc. helped to bring the Scientists, the extension personnel and the farmers much closer than before. Participation of the KAU Scientists in the programmes organised by the extension departments particularly by the Department of Agriculture is very much sought for and the KAU has always given due importance to such programmes.

Nevertheless, a matter of great concern to all those associated with the TOT process is the absence or lack of formal representation of input agencies in the statutory or non-statutory bodies of the KAU. It is true that the input agencies are invited for the State level Package of Practices Workshop and other Workshops, Seminars, *Melas*, Farmers' days etc. in which these agencies participate just like any other invited guest. The participation of the credit institutions, especially the nationalised banks and the cooperative societies is very limited. Increased agricultural production is a coordinated function involving development of improved agricultural technology, its effective transfer to the target clientele and adoption by the clientele. Rate of adoption depends mainly on the availability of inputs on time. Thus the role of input agencies in the TOT of any improved agricultural practices is inestimable. The time has come to initiate rethinking as to whether the

ISS also should be represented in the various formal linking bodies/committees, or maintain the present 'guest-host' relationship. Giving representation to the ISS in the various linking bodies will be the right step towards building a better structural linkage between and among the other sub-systems.

4.2 Extent of functional linkage between and among the research, extension, client and input sub-systems in the TOT of improved rice varieties released by KAU – results of Friedman two-way analysis of variance by ranks

4.2.1 *RSS with ESS, CSS and ISS*

In order to compare the extent of linkage between and among the research, extension, client and input sub-systems and to test the hypotheses set for the study, the scores secured by the respondents of each sub-system for the extent of linkage activities were added together. From the total, a mean linkage score for each respondent was computed and ranked. The rank totals were arranged in a two-way table. A higher rank total indicated a lesser extent of linkage activities.

To determine whether the rank totals of the sub-systems differ significantly among themselves or not, the Friedman two-way non-parametric analysis of variance test was applied. The results of analysis are presented in Table 4.1.

As it could be seen from Table 4.1, the results of the Friedman test revealed that the χ^2_r values of the RSS, ESS and CSS were significant indicating there was difference in the extent of linkage activities between and among these three sub-systems. On the other hand, the χ^2_r value of the ISS was not significant indicating that there was no significant difference in the linkage activities of the ISS with the other three sub-systems.

Further, multiple comparisons among the different sub-systems were made on the basis of sums of the ranks for extent of linkage activities and the relevant critical ranges were computed as per the procedure of Neuman (1939) and Keuls (1952) extended to ordinal data (Zar, 1981). Thus W_2 was used to compare adjacent means and W_3 to compare alternate means.

On the basis of the sums of the ranks for linkage, it could be seen from Table 4.1 that the extent of linkage of research sub-system was best with ESS (88) and this was followed by extent of linkage with

CSS (92) and ISS (132) in the descending order. It could also be inferred from the table that the RSS-ESS linkage was on par with that of the CSS. But the RSS-ISS linkage was significantly poor when compared to that of the ESS and the CSS.

TABLE 4.1: Linkages between and among the research, extension, client and input sub-systems-results of Friedman two-way non-parametric analysis of variance test

Sub-system	Sum of the ranks for linkages				χ^2_r	SE	Critical range	
	RSS	ESS	CSS	ISS			W_2	W_3
Research N = 52	—	88	92	132	22.77*	7.21	20.19	24.23
Extension N = 134	310	—	251	233	22.67*	11.57	32.41	38.89
Client N = 110-	247	156	—	257	56.31*	10.49	28.97	34.62
Input N = 32	68	59	65	—	1.31^{NS}	5.66		

* Significant at 5 per cent level of probability
NS Non-significant

Conclusions:

RSS	a* ESS	a CSS	b ISS
ESS	a ISS	a CSS	b RSS
CSS	b ESS	a RSS	a ISS
ISS	a ESS	a CSS	a RSS

* Systems having the same letter on the top do not differ significantly.

Based on these results, the general hypothesis set for the study that the extent of linkage of RSS with ESS would be on par with that of CSS and ISS (3.9.1.1) is rejected. The related hypotheses that the extent of linkage of RSS with CSS would be on par with that of ESS and ISS (3.9.1.2) and that the extent of linkage of RSS with ISS would by on par with that of ESS and CSS (3.9.1.3) are also rejected.

These results are in conformity with the observations made by Venkataraman (1988) that after the introduction of T & V system there was a well-knit and systematic linkage between research and extension sub-systems through a team of subject matter specialists.

There were 24 linkage activities which if performed by the researchers would definitely help in the TOT of improved rice varieties released by the KAU. Had all the 52 respondents performed these functions 'always' or at least 'most frequently' the sum of the ranks would have been much smaller than 84. It should also be noted that the difference in sum of the ranks between the ESS and the CSS is numerically negligible i.e. only four and is insignificant. In the case of linkage with CSS also there were 24 linkage activities which will definitely help in the TOT of improved rice varieties to the farmers. The sum of the ranks (92) indicated that the performance of the linkage activities with the CSS was still poor. This can be attributed to the fact that not all the research personnel of a station are in contact with the farmers. There is a team of scientists who were nominated as resource persons for the monthly workshops. For the joint (diagnostic) field visits, lay-out of farm trials etc. quite often these scientists go to the field and interact with the farmers. The rest of the scientists who responded to the study naturally were devoid of frequent interaction with the farmers and hence their performance of linkage activities was found to be low. The pre-monthly and monthly workshops and the combined T&V and NARP zonal workshops organised once in six months help to bring the research and the extension sub-systems closer. This must be the reason why Bhatnagar *et al.* (1986) also found that majority of the States of India indicated satisfaction regarding research-extension linkages. The result also supports the view of Pickering (1987) that there was a need for closer liaison between research and extension to ensure that research programmes are relevant to the farmers' needs.

Benor and Baxter (1984) also emphasised the necessity of undertaking joint field visits by researchers and extension workers which not only strengthen the extent of linkage between the two sub-systems but also the linkage with client sub-system. Moreover, the conduct of farm trials on farmers' fields jointly by the RSS and the ESS in fact brings these three sub-systems closer together and hence the result.

The poor or less effective linkage between the RSS and the ISS can be attributed to the fact that these two sub-systems often come into close contact only on special occasions like seminars, melas and field days. There is no well-knit and regular linkage mechanism between these sub-systems as in the case of the RSS with the ESS. Linkage activities like conducting trials with pesticides, fertilisers etc. by the

RSS are now on a limited scale. Similarly, involvement of the ISS in village adoption programme, lab-to-land programme etc. is not up to the mark. The need for effective linkage between and among the RSS, ESS, CSS and ISS was also stressed by Jaiswal and Arya (1981), Reddy (1981), Sharma (1982), Sen (1984), Kishore (1986) and Bhatnagar and Desai (1987). Of these, the observations of Kishore (1986) appear to be more relevant in this context. Suggesting an alternate model for effective TOT, Kishore underscored the exigency of strengthening the linkage with input systems which, according to him, was the weakest link in the chain of TOT.

4.2.2 *ESS with RSS, CSS and ISS*

The extent of linkage of ESS was found to be the best with ISS (233) and this was followed by extent of linkage with CSS and RSS (251 and 310 respectively) in the descending order (Table 4.1). It could also be seen that the ESS-ISS linkage was on par with that of ESS-CSS linkage. On the contrary, the ESS-RSS linkage was not on par with that of CSS and ISS. In the light of these findings, the general hypothesis set for the study that the extent of linkage of ESS with RSS would be on par with that of CSS and ISS (3.9.1.4) is rejected. The other two related hypotheses viz. the extent of linkage of ESS with CSS would be on par with that of RSS and ISS (3.9.1.5) and the extent of linkage of ESS with ISS would be on par with that of RSS and CSS (3.9.1.6) are also not sustainable and hence rejected.

It is in fact intriguing to find that when the RSS claimed to have the best linkage with ESS, the ESS was found to have least effective linkage with RSS. These contradictory views of the two sub-systems could be attributed to the fact that RSS is in regular contact with the ESS during pre-monthly and monthly workshops, zonal workshops of T & V system and NARP and during diagnostic field visits. And these workshops are organised at the research stations where there is ample scope to interact with each other.

On the contrary, in the case of the ESS, only the higher level officers (the SMS, SDAOs and PAO of the district) attend the pre-monthly workshops, zonal workshops and diagnostic field visits. The respondents of the ESS included both higher level officers and the lower category of officers like AOs and ADs working at the AE unit and village level. Majority (77.88%) of the respondents of ESS were AOs and ADs who very rarely come into direct contact with the RSS and naturally there is lesser linkage between these category of officers and research personnel.

The results of the study are in agreement with the observations of Venkataraman (1988) that after the introduction of the T & V system, there was a well-knit extension-research linkage through a team of SMS. But both the extension and research sub-systems should learn to work as a single unit to render effective service to the farmers.

The findings of Malekmohammadi (1987) that researchers in agriculture need creative relations with extension workers and also agricultural educators and *vice versa*, in order to activate extension movement and at the same time research and education activities in agriculture particularly holds good in this context.

These results evidently exhort to streamline the ESS-RSS linkage. This could possibly be achieved by giving due representation to AOs and ADs also in the monthly and zonal workshops so that these officers could also interact directly with the researchers and reflect the field problems as perceived by them.

Sum of the ranks for linkage of ESS also indicated that the ESS had maximum extent of linkage with the ISS and this was very closely followed by linkage with the CSS. This shows the well-knit relationship existing between the ESS and the ISS particularly after the introduction of T & V system. The inputs requirement of the farmers are assessed by the ISS through the ESS based on the fortnightly messages and naturally, the ESS-ISS linkage has considerably improved.

The better ESS-CSS linkage could be explained in view of the regular and scheduled farm visits to the contact and other farmers by the extension personnel under the T &V system. Once the extension service in the State became strong, it helped to generate a larger demand for input supplies and credit and these helped gear up the activities of input agencies and consequently enhanced the volume of business and credibility of these agencies. Thus the synergistic influence of the CSS and the ISS helped to keep the ESS-CSS linkage almost on par with ESS-ISS linkage.

4.2.3 *CSS with RSS, ESS and ISS*

As it is evident from the Table 4.1 linkage of the CSS was highest with the ESS (156) and this was followed by RSS and ISS (247 and 257, respectively) in the descending order. Further the CSS-RSS linkage was statistically on par with that of the ISS. But the extent of linkage of the CSS with the ESS was significantly better than that with the

RSS and the ISS. In the light of the above results, the general hypothesis that the extent of linkage of CSS with RSS would be on par with that of ESS and ISS (3.9.1.7) is rejected. The related hypotheses that the extent of linkage of CSS with ESS would be on par that of RSS and ISS (3.9.1.8) and that the extent of linkage of CSS with ISS would be on par with that of RSS and ESS (3.9.1.9) are also rejected.

As evidenced from the table, the CSS had performed the maximum extent of linkage activities with the ESS because of the time-bound fortnightly farm visits by extension personnel under the T & V system. It would be worth recalling that the ESS also had the second highest linkage with CSS. This type of mutual agreement on the extent of linkage between the two sub-systems would be quite encouraging and conducive for TOT of improved rice varieties.

Nevertheless, the fact remains that the CSS-RSS linkage and CSS-ISS linkage were not that encouraging and hence warrant further strengthening for effective TOT process. The linkage between the RSS and CSS to be effective, there must be a regular feedback from the CSS to the RSS through the ESS (Rogers, 1983). But in the present set up, the linkage of the RSS with the CSS is confined to diagnostic field visits, farm trials on the farmers' fields and stray contacts during the seminars, field days etc. Hence, urgent steps need to be initiated by the RSS to strengthen its linkage with the CSS further through on farm training of farmers and other extension activities.

The ISS is now found to have very little direct linking functions with the CSS. The material inputs like fertilizers, pesticides etc. are made available mainly through a myriad authorised dealers whose motive is only profit-making. For effective diffusion of the technology, the ISS should have a battery of sales points at reasonable distances with adequate consultancy and service facilities. Matlock (1985) has rightly pointed out that ISS was the side-support in the framework of agricultural production without which even the best technology would prove barren, abortive and abstract. The suggestion of Kishore (1986) for better coordination between and among KCS, KDS and KCS including the ISAS is quite relevant in this context and is in line with the results of the present study.

4.2.4 *ISS with RSS, ESS and CSS*

Data in Table 4.1 revealed that the extent of linkage of the ISS was the highest with the ESS (59) and this was closely followed by the CSS (65) and the RSS (68). The χ_r^2 value of the sums of the ranks was not significant indicating that there was no significant difference in the linkage activities of the ISS with every other sub-system. Therefore, the general hypothesis set for the study – the extent of linkage of ISS with RSS would be on par with that of ESS and CSS (3.9.1.10) is not rejected. The related hypothese – the extent of linkage of ISS with ESS would be on par with that of RSS and CSS (3.9.1.11) and the extent of linkage of ISS with CSS would be on par with that of RSS and ESS (3.9.1.12) are also acceptable.

It is interesting to note that in the case of ESS-ISS linkage, the ESS had the maximum extent of linkage with the ISS. Now, the ISS has expressed its reciprocal linkage with the ESS which indicates a perfect mutual agreement in its functional relationship. On the basis of these findings, it can be concluded that there existed a well established functional relationship between the ISS and the ESS which was conducive for effective TOT process.

Similarly, the ISS-CSS linkage and ISS-RSS linkages are almost on par with each other, the sum of the ranks for linkage being approximately equal (65 and 68, respectively — Table 4.1). It could also be seen that these results are in perfect consonance with CSS-ISS and RSS-ISS linkages, in all the cases the linkages being less effective and least effective.

The crucial point emerging from the study which causes concern in the TOT process is the fact that even after five years of implementation of the T & V system in the State, there is still lack of coordination between and among the ESS-RSS, RSS-ISS, CSS-RSS and CSS-ISS, among the various combinations analysed.

The results of the study on the extent of linkages between and among research, extension, client and input sub-systems coalesce together in conformity with the observations of Kishore (1986) that there is an urgent need for an alternate strategy for the TOT in the form of an integrated approach and closer cooperation between and among KGS, KDS, KCS and ISAS.

The foregoing findings and discussions lead to corroborate the following conclusions with regard to the extent of linkages between and among the research, extension, client and input sub-systems in the TOT of improved rice varieties released by the KAU.

1. There were significant differences in the extent of linkage between the RSS-ISS, ESS-RSS, CSS-RSS and CSS-ISS.
2. Though the RSS opined to have effective linkage with ESS, the ESS did not subscribe to this view. Hence there is a need to strengthen the ESS-RSS linkage by way of involving the AOs and ADs also in the monthly workshops and zonal workshops of

T & V system and NARP.

3. The RSS-CSS linkage was also found to be not eftective. Though the RSS claimed to have effective linkage with the CSS, the CSS did not subscribe to this view and hence this needs strengthening. The RSS-CSS linkage could possibly be strengthened by interacting the two sub-systems more frequently than at present.
4. The RSS-ISS linkage was found to be not up to the mark and there was mutual agreement of the two sub-systems in support of this view. Hence, both the sub-systems should evolve ways and means of improving the linkages between the two. The more feasible proposition seems that the ISS increases its frequency of visits to research stations and taking the help of researchers in the field extension activities of the ISS.
5. The CSS-ISS linkage was found to be not satisfactory and both the sub-systems contributed to this view. The ISS, therefore should evolve appropriate strategy to improve its linkage with CSS who are the major consumers/utilizers of the inputs of supplies and credit.

   The CSS look down the ISS as mere salesmen functioning with prime motive of making profit. Opening more number of sales points and service centres by the input supplying agency directly at reasonable distances would perhaps help to increase the credibility of the ISS and thereby augment the linkage between the two sub-systems. The credit agencies are very rarely considered by the CSS as an extension agency, for, the officers are not in a position to impart any technical advice to the farmers because the agricultural officers working in the nationalised banks are completely devoid of field exposure.
6. The ESS-CSS and the ESS-ISS linkages are found to be satisfactory and all the three sub-systems mutually agree to this view. The situation is conducive for effective TOT. The situation needs to be exploited by integrating and strengthening the RSS-ESS, RSS-ISS, RSS-CSS and CSS-ISS linkages. The analysis also brings to limelight yet another significant finding. In all cases, excepting the ISS, the hypotheses that the extent of linkage activities between and among the sub-systems would be on par have been rejected. This indicates that, there was generally a preferential pattern of linkage functions leading to the near neglect of some sub-systems. It appears that the concept of systems approach in TOT process has not percolated through to the various sub-systems involved in the TOT process, which requires immediate amelioration.

TABLE 4.2: Principal-factor solutions for extent of linkage of research sub-system with extension sub-system—factor patterns and their loadings

	Activities	Factor							Communality (h^2)
		I	II	III	IV	V	VI	VII	
X_1	Involvement in training extension personnel	.4606	.1867	.2485	.1790	.0566	.5020	.1566	.6206
X_2	Serving as course director of training	.1762	–.1150	.4261	.2674	.3156	.6494	–.1827	.8520
X_3	Serving as resource person of training	.6814	.1284	–.0218	.2382	–.0283	.2605	–.2242	.6568
X_4	Serving as consultant for organizing training	.5279	.1626	.2176	.3615	–.1118	.1527	–.0918	.5275
X_5	Serving as resource person for diagnosing field problems	.8008	.1065	.2926	.0835	–.0969	–.0917	–.1781	.7947
X_6	Participation in joint field visit	.7222	.0658	.3485	.0535	–.1549	–.1015	–.1170	.6982
X_7	Serving as resource person on field days	.4609	.4737	.3099	.3710	.0848	.0399	–.1971	.7177
X_8	Serving as resource person for farmers' days	.4408	.2581	.3109	.4504	.1984	.0342	–.3866	.7504
X_9	Serving as resource person in monthly workshops	.5484	–.1594	.1506	.0035	–.2376	.1534	–.3581	.5571
X_{10}	Attending regional workshops of NARP and T & V system	.0458	–.1315	–.0880	–.2712	.1286	–.1097	–.2054	.1714
X_{11}	Involvement as implementing officer of lab-to-land programme	–.0033	–.5572	.2449	.5559	–.0676	.0733	–.1222	.7044

(Contd.)

(Contd.)

	Activities	Factor							Communality (h^2)
		I	II	III	IV	V	VI	VII	
X_{12}	Serving as implementing officer of village adopting programme	.0400	–.6944	.0782	.5779	–.3037	.0109	–.0843	.9233
X_{13}	Participation in conducting farm trials	.4340	–.1441	.4060	–.0008	–.3534	.2539	–.3309	.6728
X_{14}	Visiting result demonstrations	.1815	–.0177	.9674	–.1113	–.0852	–.0463	.0250	.9915
X_{15}	Visits to minikit trials	.2074	.2105	.7286	.0649	–.2302	.0767	–.1228	.6963
X_{16}	Participation in exhibitions	.3383	.2959	–.0425	.2217	–.1193	–.0578	–.1155	.2839
X_{17}	Participation in extension seminars	.5536	.3280	.0980	.3650	–.3251	–.0923	.0388	.6725
X_{18}	Participation in extension workshops	.3517	.4300	.0608	.5872	–.3064	–.0013	–.1593	.7764
X_{19}	Writing in popular publications	.1207	–.0015	.1787	.2528	–.6509	.0304	–.0541	.5379
X_{20}	Writing in newspaper columns	.1103	–.0701	.2407	.0574	–.5295	–.4420	–.0977	.5636
X_{21}	Attending package of practices workshop	.3728	–.0308	.1237	–.0049	–.3349	.1441	–.8457	.9999
X_{22}	Attending minipackage workshops	.3279	–.0054	.0907	.1493	–.3587	–.0001	–.7564	.8388
X_{23}	Participation in KAU rice variety release committee	.1924	–.0715	.0459	.0653	–.5257	.1014	–.3909	.4879
X_{24}	Participation in state rice variety release committee	.1047	.2182	–.1722	.0887	–.5930	.0654	.1380	.4712
	Proportionate variance accounted by each factor	.1663	.0724	.1075	.0835	.0974	.0476	.908	

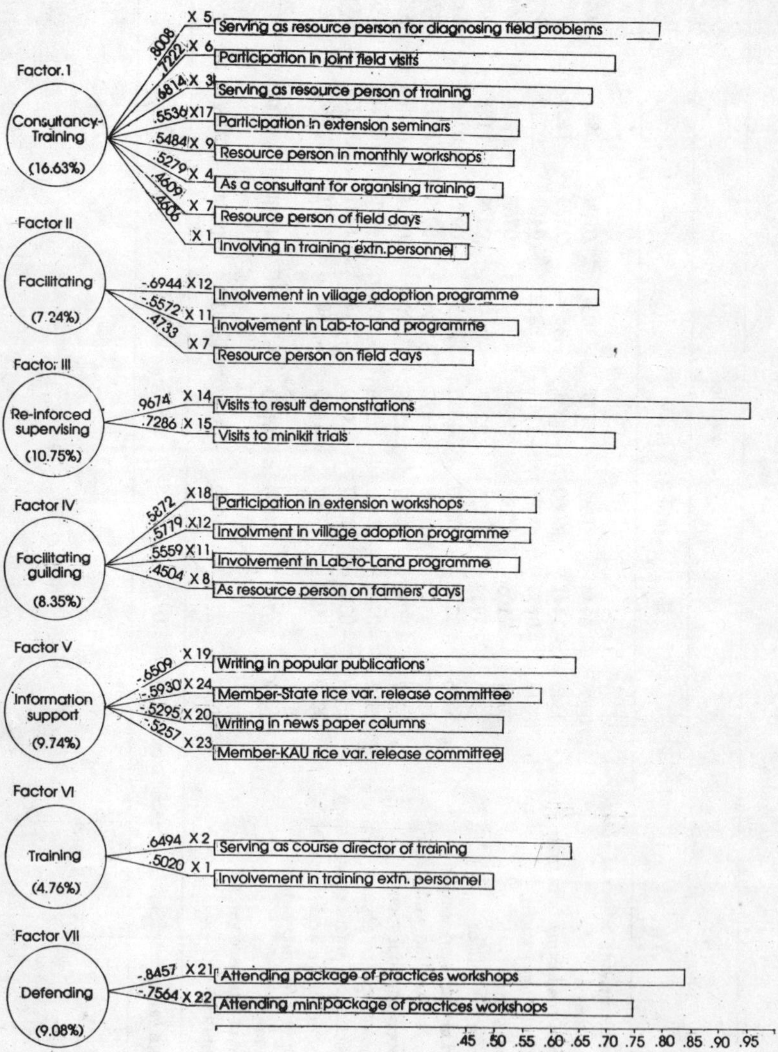

FIG. 4.4: **Diagrammatic representation of principal-factor** solutions for RSS - ESS **linkage**

4.3 Factors associated with TOT of improved rice varieties released by the KAU in respect of research, extension, client and input sub-systems-results of principal-factor analysis

With a view to segregate the factors influencing the extent of linkage between and among the four sub-systems, factor analysis was done using principal-factor method. The factors extracted were appropriately labelled and the results are furnished in this section.

4.3.1 *RSS-ESS Linkage*

The Principal-factor solutions for the extent of linkage of RSS with ESS are presented in Table 4.2. The results clearly indicate that about 66 per cent of the total variability in the extent of linkage of the RSS with the ESS was accounted for by seven factors.

The content and nature of individual factors are described below:

Factor I

The proportionate contribution of Factor I to the total variability was 16.63 per cent which is the highest of the seven factors. As many as eight linkage activities under Factor I were found to have relatively high positive loadings well above the limit of 0.45 (Walton, 1972). Out of these, linkage activity X_5-serving as resource person for diagnosing field problems was found to have a loading of .8008 which according to Fruchter (1967) is very high. This was followed by X_6-participation in joint field visits (.7222), X_3-serving as resource person of training (.6814), X_{17}-participation in extension seminars (.5536), X_9-serving as resource person in monthly workshops (.5484) and X_4-serving as a consultant for organising training (.5279). Linkage activity X_7 and X_1 had only moderate loadings of .4609 and .4606 and these activities again appear with high loadings under Factors II and VI respectively.

The correlation between any two activities under Factor I was positive and significant indicating their interrelationship. These activities come under the broad nature of the characteristics of 'consultant' and 'trainer' identified by Havelock *et al.* (1969) and therefore with reasonable conviction, Factor I could be labelled as 'consultancy-training'.

Factor II

This factor contributed 7.24 per cent to the total variability. Only three linkage activities were found to have significant loadings under Factor II. They were X_{12}-involvement as implementing officer of village adoption programme (–.6944), X_{11}-involvement as implementing officer of lab-to-land programme (–.5572) and X_7-serving as resource person on field days (.4737). As it could be seen, linkage activities X_{12} and X_{11} were negatively correlated but highly loaded and only X_7 was positively correlated, though only moderately loaded. According to Harman (1960) bi-polar factors are not essentially different from any other but is merely one for which several of the variables have significant negative projections.

The linkage activities of implementing lab-to-land and village adoption programmes and serving as resource person on field days are essentially 'facilitating' functions and, therefore, Factor II could reasonably be termed as 'Facilitating'.

Factor III

The proportionate contribution of this factor to the total variability was 10.75 per cent being the highest contributor after Factor I. Factor III was mainly influenced by two linkage activities viz. X_{14}-visits to result demonstrations (.9674) and X_{15}-visits to minikit trials (.7286). Both the activities were very highly loaded and significantly correlated. These two linkage activities are performed by the RSS with the objective of reinforcing the results obtained in research stations and to supervise their performance in field situation. Therefore, Factor III could logically be labelled as "Reinforced Supervising".

Though these two activities are mainly the domain of extension personnel, involvement of farm scientists will contribute to a better integration of research and extension activities as extension is the process of extending the research findings to the cultivators' fields. Therefore, a close liaison must exist between those who are responsible for inventing and innovating and those who extend the results to the farmers. Mirdha (1974) has rightly pointed out that the extent to which the process of modernisation of agriculture takes momentum depends upon the integration of research and extension and the extent to which those engaged in these two functions have mutual understanding and collaboration at various levels.

Factor IV

This factor contributed 8.35 per cent to the total variability of linkage of RSS to ESS. Four linkage activities were found to have significant loading under this factor. These included X_{18}-participation in extension workshops (.5872), X_{12}-involvement as implementing officer of village adoption programme (.5779), X_{11}-involvement as implementing officer of lab-to-land programme (.5559) and X_8-serving as a resource person on farmers' day celebrations (.4504). Linkage activities X_{12} and X_{11} were also found to be contributing to Factor II but negatively correlated. Activity X_8 is just at the limit of 0.45, but positively correlated.

On analysis of the linkage activities under this factor, it becomes clear that besides the 'facilitating' function identified under Factor II, the function of 'guiding' the extension workshops and farmers' day celebrations is also involved in this factor and hence it is labelled as 'facilitating-guiding'.

Factor V

The contribution of Factor V to total variability was 9.74 per cent. The linkage activity X_{19}-writing in popular publications had the highest factor loading of –.6509 under this factor. This was followed by X_{24}-participation in State rice variety release committee (–.5930), X_{20}-writing in Newspaper column (–.5295) and X_{23}-participation in KAU rice variety release committee (–.5257). All the four activities, though with high loadings, were negatively correlated. Activities X_{19} and X_{20} are essentially 'communication' functions and participation in State rice variety release committee (X_{24}) and KAU rice variety release committee (X_{23}) is to provide knowledge on the rice varieties released by KAU. Therefore, Factor V could be termed as 'Information support'.

Factor VI

The proportionate contribution of Factor VI to the total variability was only 4.76 per cent. Results of varimax rotation indicate that only two linkage activities were significantly loaded under Factor VI. They were X_2-serving as course director of training (.6494) and X_1-involving in training extension personnel (.5020). Both the activities which constitute Factor VI are positively correlated and since these relate to training, this factor could be obviously termed as "training".

Factor VII

Varimax rotation of the factor matrix revealed that only two linkage activities X_{21}-attending package of practice workshops (–.8457) and X_{22}-attending minipackage workshops (–.7564) were mainly contributing to Factor VII. The proportionate contribution of variability by the factor was 9.08 per cent. Though negatively correlated, both activities have very high loadings. The common name coined for this factor is 'defending' since in these workshops, the researchers are expected to defend and convince the other participants about the new rice varieties and technologies released by the KAU.

4.3.2 *RSS-CSS Linkage*

Principal-factor solutions for RSS-CSS linkage and their factor loadings are presented in Table 4.3. A total of six factors had been extracted as a result of varimax rotation of factor matrix. The total variability accounted for by the six factors to linkage of RSS to CSS was 56.00 per cent.

The content and nature of individual factors are described below:

Factor I

This factor had a contribution of 17.62 per cent to the total variability of the hypothetical construct. Factor I was dominated by eight linkage activities having significant factor loadings well above the limit of 0.45. The activities included X_{24}-participation in group discussions organised by farmers (.8062), X_{22}-organising exhibitions in research stations (.7102), X_{23}-organising group discussions of farmers (.6859), X_{18}-liaison with cooperative societies of the area (.6181), X_{21}-participation in campaigns (.6145) X_{10}-clarification of doubts of farmers through letters (.6038), X_{1}-organising training for farmers (.5542) and X_{19}-liaison with farmers' association of the area (.5379) in the descending order of factor loadings. All the activities had fairly high factor loadings and positively correlated.

TABLE 4.3: Principal-factor solutions for extent of linkage of research sub-system with client sub-system—factor patterns and their loadings

	Activities	Factor						Communality (h^2)
		I	II	III	IV	V	VI	
X_1	Organizing training for farmers	.5542	.1419	–.1475	–.0079	.2420	.8030	.4243
X_2	Involvement in training as resource person	.4324	.2379	.0547	–.0955	.2214	.0800	.3123
X_3	Visits to farm trials	.4190	.4177	.2048	–.1042	.1111	.4660	.7031
X_4	Contribution of articles in newspaper columns	.1914	.4559	–.2420	.0836	.3445	.0858	.4443
X_5	Contribution of articles in magazines	.0716	.3995	–.2761	.1974	.4556	–.1089	.5171
X_6	Contribution of articles in souvenirs	.0877	.5850	–.2027	–.0604	.1950	.2028	.5538
X_7	Contribution of articles in books for farmers	1609	.2496	–.0366	.0285	.5201	.0458	.3632
X_8	Participation in radio programmes for farmers	.3725	.1856	–.3844	.0121	.3350	–.2182	.4809
X_9	Clarification of doubts of farmers on rice in person	.2961	.0826	–.5523	–.5128	–.1249	–.1910	.7216
X_{10}	Clarification of doubts of farmers on rice through letters	.6038	.2678	–.2273	–.1187	.0969	.0227	.5374
X_{11}	Involvement in organizing farmers days	.1297	–.0790	–.7335	.3826	–.1012	.0078	.7400
X_{12}	Involvement in organizing field days	.3010	–.0907	–.6112	.4159	.0126	.1193	.6777
X_{13}	Visits to farmers fields on request	.1236	.1972	–.5333	–.1928	–.0790	.0589	.4004
X_{14}	Taking the visiting farmers to experimental plots	.0086	–.0714	–.6406	–.2798	.3476	.0310	.6874
X_{15}	Serving in the regional inter-disciplinary committee	.0373	.7915	–.2528	–.0118	–.1294	–.1684	.7752
X_{16}	Serving in the advisory committee on national demonstration	–.0117	.0126	–.0493	–.0610	–.0152	.0630	.1866
X_{17}	Involvement in front line demonstration	.2800	.2716	–.0709	–.0298	.0098	–.3753	.3156
X_{18}	Liaison with co-operative societies of the area	.6181	–.0023	–.3240	–.3739	.0671	.2511	.8791
X_{19}	Liaison with farmers associations of the area	.5379	.0592	–.2740	–.2115	.4699	–.1692	.7245
X_{20}	Contribution of lessons for correspondence course on rice	–.0458	.6486	.0503	–.0612	.1123	–.0770	.4479
X_{21}	Participation in campaigns	.6145	.1654	–.0607	.0536	.0559	.1221	.4533
X_{22}	Organizing exhibitions in research stations	.7102	.0516	.0734	.0997	.1363	–.0765	.5473
X_{23}	Organizing group discussions of farmers	.6859	.0165	–.2478	.1067	.1025	.0741	.5880
X_{24}	Participation in group discussions organized by farmers	.8062	.1702	–.2594	.0311	.0341	–.3158	.9227
	Proportionate variance accounted by each factor	.1762	.1002	.1244	.0461	.0652	.0492	

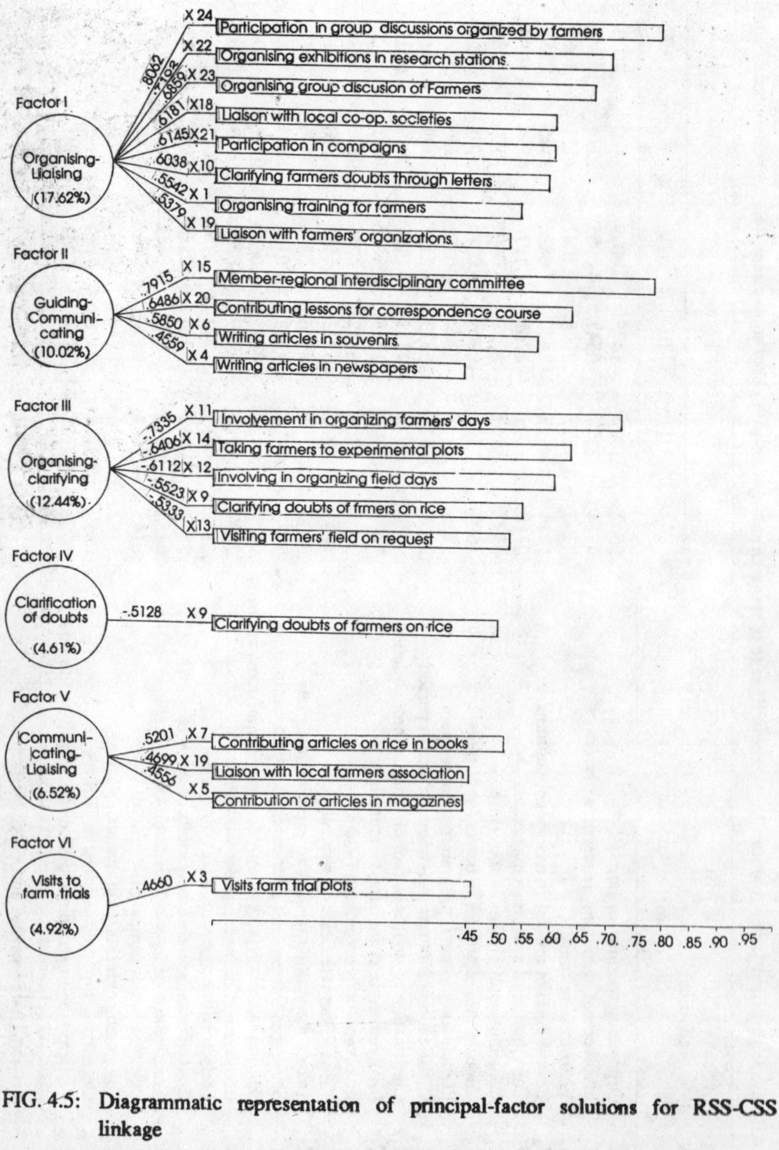

FIG. 4.5: Diagrammatic representation of principal-factor solutions for RSS-CSS linkage

A detailed analysis of the Factor I indicates the need for getting the researchers out of the research stations and associate themselves actively with farmers and helping to solve their problems. Although research system is mainly concerned with the creation of new technology, it is not an end in itself. The ultimate object is the widespread adoption of the new varieties or technologies by the farming community. To achieve this end, the research sub-system can link itself with clients by devoting part of their time in organising training for farmers, conducting discussions, campaigns, organising exhibitions and establishing liaison with farmers organisations and institutions. Similar view has been expressed by Havelock *et al.* (1969) saying that knowledge production and knowledge use represent two systems, each having its own characteristic-sub-cultural norms, values, communication patterns etc. The rationale for a link between the knowledge production system (research sub-system) and knowledge use system (client sub-system) is that scientists and farmers are the products of two distinct 'worlds'. This implies that when left to their own device, the two sub-systems may not 'meet' each other and the new rice varieties and their technology will become intangible for the client system. Venkataraman (1988) also exhorted the scientists to increase the field visits, and the scientists and research stations should become more visible and readily accessible to the farmers as well as the extension workers.

Factor I can be termed as 'organising-liaising' since majority of the activities under this factor reflect these functions performed by the RSS.

Factor II

The proportionate contribution of variability by Factor II was 10.02 per cent. This factor had four linkage activities with significant factor loadings. They were X_{15}-serving as member of regional inter-disciplinary committee (.7915), X_{20}-contribution of lessons for correspondence course on rice (.6486), X_6-writing articles in souvenirs on rice (.5850) and X_4-writing articles on rice in newspapers (.4559). X_{15}, X_{20} and X_6 had fairly high loadings and all the four activities were positively correlated. With reasonable conviction, this could be described as "Guiding-Communicating" factor.

Factor III

This factor had a contribution of 12.44 per cent to the total variability. The factor was mainly influenced by five linkage activities viz. X_{11}-involvement in organising farmers' days (–.7335), X_{14}-taking farmers to experimental plots (–.6406), X_{12}-involving in organising rice day (–.6112), X_9-clarifying doubts of farmers on rice in person (–.5523) and X_{13}-visits to farmers' fields on request (–.5333). All the five activities had high loadings but were negatively correlated. All the five activities put together can be termed as "organising-clarifying" factor.

Factor IV

This factor contributed only 4.61 per cent variability to the hypothetical construct-linkage of RSS with CSS. Only one linkage activity viz. X_9-clarification of doubts of farmers on rice in person was having significant loading but it was negatively correlated (–.5128). This activity was present under Factor III too but there it was positively correlated.

Factor V

Factor V had a contribution of 6.52 per cent variability. There were only three linkage activities under this factor viz. X_7-contribution of articles on rice in books for farmers (.5201) and X_{19}-liaison with farmers' associations of the area (.4699) and X_5-contribution of articles in magazine (.4556). All the activities were positively correlated and could be described as "communicating-liaising" factor.

Factor VI

The lone linkage activity under Factor VI was X_3-visits to farm trials (.4660) which had a contribution of 4.92 per cent to the total variability.

Farm trials are conducted by the extension personnel with the technical support of research system on the farmer's fields. This activity gives ample scope for scientist-farmer interaction.

4.3.3 *RSS-ISS Linkage*

Principal-factor solutions for RSS-ISS linkage are presented in Table 4.4. In all, six factors had been extracted which together contributed to 61.70 per cent variability. The content and nature of individual factors are given below:

Factor I

The proportionate variability contributed by Factor I was 19.52 per cent. As many as seven linkage activities were found to have significant loadings well above the limit of 0.45 under Factor I. Taking individual linkage activities into consideration; X_8-inviting weedicide dealers in seminars had very high factor loading (.8567). This was followed by X_9-inviting pesticide dealers in seminars (.8000), X_{10}-inviting credit agencies in seminars (.6993), X_{18}-involving input agencies in exhibitions (.6945), X_7-inviting fertilizer dealers in seminars (.6354), X_{20}-involving input agencies in national demonstrations (.5389) and X_{19}-involving input agencies in farm trials (.4652) in the descending order of factor loadings. Considering the nature of the linkage activities clustering under Factor I, this factor could be designated as "enabling".

Factor II

This factor contributed 12.62 per cent variability to the hypothetical construct. There were six linkage activities having loadings above the minimum limit of 0.45. These included X_5-conducting trials with fertilizers supplied by dealers (–.7482), X_{15}-helping fertilizer dealers in laying out demonstrations (–.7187), X_{13}-feeding back the results of fertiliser trials to dealers (–.6283), X_{12}-participation in workshops organised by input agencies (–.5621), X_{20}-involving input agencies in national demonstrations (–.5227) and X_{21}-contributing articles in publications of input agencies (–.4920). All the activities were significantly but negatively loaded. Of these, activity X_{20} appeared under Factor I also with positive loading. All the linkage activities coming under this factor could be named collectively as "facilitating-communicating".

TABLE 4.4: Principal-factor solutions for extent of linkage of research sub-system with input sub-system—factor patterns and their loadings

	Activities	Factor						Communality (h^2)
		I	II	III	IV	V	VI	
X_1	Keeping contact with fertilizer dealers	.2432	–.1475	.1987	–.2589	–.1121	.5432	.4950
X_2	Keeping contact with pesticides dealers	.3483	.0984	.4638	–.4650	–.1952	.3515	.7242
X_3	Keeping contact with local nationalised banks	.0515	–.2983	.0030	–.1283	–.6883	–.1334	.5996
X_4	Keeping contact with local co-operative societies	.3269	.1074	.1884	–.1665	–.7714	.1809	.8095
X_5	Conducting trials with fertilizers supplied by dealers	.1242	–.7482	–.0763	–.0417	–.0719	.1656	.6154
X_6	Conducting trials with pesticides supplied by dealers	.4442	–.0810	.3526	–.3830	–.0948	–.0559	.4869
X_7	Inviting fertilizer dealers for seminars at res. stations	.6354	.0370	.0008	–.0798	–.0684	.5826	.7557
X_8	Inviting weedicide dealers for seminars at res. stations	.8567	–.1470	.0098	–.0410	–.0468	.2330	.8137
X_9	Inviting pesticide dealers for seminars at res. stations	.8000	–.0597	.1055	–.2008	–.0393	.2737	.7713
X_{10}	Inviting credit agencies for seminars at res. stations	.6993	.0124	.1749	–.2064	–.2195	.1516	.6335
X_{11}	Participation in seminars organized by input agencies	.2703	–.3343	.4858	–.3848	–.1004	.2102	.6232
X_{12}	Participation in workshops organized by input agencies	.3270	–.5621	.3123	–.1646	–.0403	.0916	.5575
X_{13}	Feeding back the results of fertilizer trials to dealers	.0675	–.6283	–.0419	–.1474	–.0984	.5265	.6994
X_{14}	Feeding back the results of pesticide trials to dealers	.1478	–.1704	.2500	–.6919	–.0418	.2323	.6565
X_{15}	Helping fertilizer dealers in layingout demonstrations	.4181	–.7187	–.0242	–.0652	.0634	.0745	.7057
X_{16}	Involving input agencies in lab-to-land programme	.2329	–.0356	–.4712	–.5219	–.1095	.0337	.5631
X_{17}	Involving input agencies in village adoption programme	.2724	–.0982	–.6848	–.6110	–.0189	.1085	.9378
X_{18}	Involving input agencies in exhibitions	.6945	–.1540	–.1307	–.1614	–.3234	.2275	.7055
X_{19}	Involving input agencies in farm trials	.4652	–.2684	–.1994	–.1603	–.1805	.0188	.3869
X_{20}	Involving input agencies in national demonstrations	.5389	–.5227	–.1392	.1434	.0424	–.2188	.6533
X_{21}	Contributing articles in publications of input agencies	.1409	–.4920	.1198	–.1615	–.0155	–.0769	.3085
X_{22}	Helping input agencies as consultants	.3781	–.4075	.2788	–.4619	–.1855	–.1696	.6632
X_{23}	Maintaining contact with input agencies through correspondence	.2216	–.1456	.0034	–.5984	–.3081	.1340	.5411
	Proportionate variance accounted by each factor	.1952	.1262	.0748	.1109	.0651	.0672	

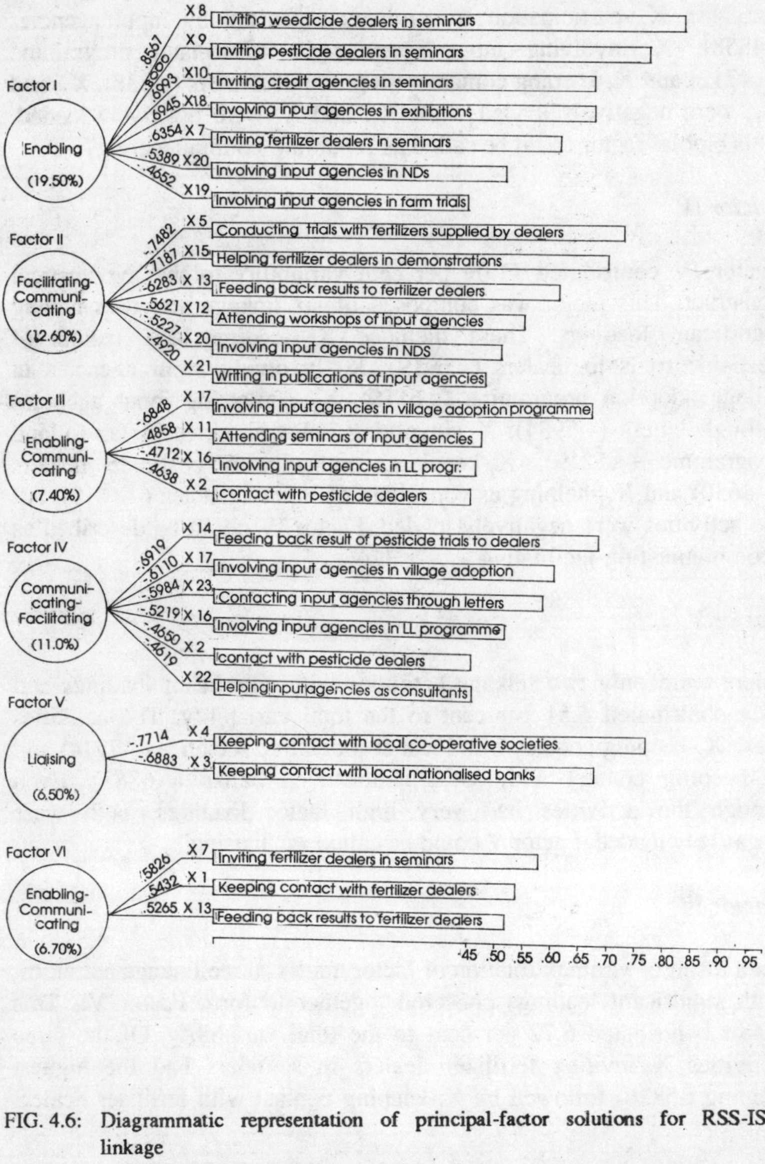

FIG. 4.6: Diagrammatic representation of principal-factor solutions for RSS-ISS linkage

Factor III

The contribution of Factor III to the total variability was 7.48 per cent. Out of 23 linkage activities only four had significant factor loadings. They were X_{17}-involving input agencies in village adoption programme (–.6848), X_{11}-participation in seminars organised by input agencies (.4858), X_{16}-involving input agencies in lab-to-land programme (–.4712) and X_2-keeping contact with pesticide dealers (.4638). X_{17} and X_{16} were negatively loaded while X_{11} and X_2 were positively loaded. This bipolar factor could be called as "enabling-communicating".

Factor IV

Factor IV contributed 11.09 per cent variability to the hypothetical construct. This factor was composed of six linkage activities having significant loadings. These included X_{14}-feeding back results of pesticide trials to dealers (–.6919), X_{17}-involving input agencies in village adoption programme (–.6110), X_{23}-contacting input agencies through letters (–.5984), X_{16}-involving input agencies in lab-to-land programme (–.5219), X_2-keeping contact with pesticide dealers (–.4650) and X_{22}-helping as consultant of input agencies (–.4619). All the activities were negatively loaded. Factor IV could be described as "communicating-facilitating".

Factor V

There were only two linkage activities with significant loadings and they contributed 6.51 per cent to the total variability. The activities were X_4-keeping contact with local co-operative societies (–.7714) and X_3-keeping contact with local nationalised banks (–.6883). Even though the activities had very high factor loadings, both were negatively loaded. Factor V could be called as "liaising".

Factor VI

As a result of varimax rotation of factor matrix, three linkage activities, with significant loadings clustered together to form Factor VI. This factor contributed 6.72 per cent to the total variability. Of the three activities X_7-inviting fertilizer dealers in seminars had the highest loading (.5826) followed by X_1-keeping contact with fertiliser dealers

(.5432) and X_{13}-feeding back results of fertiliser trials to dealers (.5265). All the three linkage activities pooled together can be called as "enabling-communicating". These activities go parallel with those identified under Factor III which is indicative of the importance of 'enabling-communicating' function of the RSS performed with the ISS in the TOT of rice varieties released by the KAU.

4.3.4 *ESS-RSS Linkage*

Principal-factor solutions for ESS-RSS linkage are presented in Table 4.5. As it could be seen from the table varimax rotation extracted five factors from the 21 linkage activities. All the five factors put together the total contribution to the variability was 50.80 per cent.

The nature and content of individual factors are described below:

Factor I

Among the individual factors, the first factor is the maximum contributor towards the total variability (14.74%). There were four linkage activities with significant loadings under this factor. Activity X_4-participation in monthly workshops of T & V system had the highest loading (.8997). This was followed by X_5-participation in zonal workshops of NARP and T & V system (.7361), X_3-participation in pre-monthly workshops (.5493) and X_7-participation in joint field visits with researchers (.4728). All the four linkage activities pooled together could be called as "knowledge building and communicating" factor.

As stated elsewhere, an extension service cannot function in isolation and there should be frequent interaction between the extension and research sub-systems. The established step towards this type of functional relationship between the two sub-systems is the pre-monthly and monthly workshops under the T & V system, the participation in combined NARP and T & V zonal workshops and participation in joint field visits with researchers. The extension personnel and scientists are brought together periodically on such forums and they exchange field problems and solutions. This also builds the level of knowledge of the extension personnel who in turn feedback the relevant information to the researchers for further investigation.

TABLE 4.5: Principal-factor solutions for extent of linkage of extension sub-system with research sub-system—factor patterns and their loadings

	Activities	Factor					Communality (h^2)
		I	II	III	IV	V	
X_1	Visiting experimental plots at rice res. stations	.1827	.1153	.2957	.1531	.1734	.3866
X_2	Co-operating with researchers in layingout farm trials	.1179	–.2283	.2328	.5033	.1885	.4238
X_3	Participation in pre-monthly workshops of T & V system	.5493	–.0168	.1449	.1876	.3434	.5275
X_4	Participation in monthly workshops of T & V system	.8997	.1653	.2300	.2231	.0098	.9398
X_5	Participation in zonal workshops of NARP & T & V system	.7361	.1372	.1406	.0322	.1178	.6026
X_6	Taking field problems on rice to researchers	.2462	–.1441	.1191	.4326	.2591	.3671
X_7	Participation in joint field visits with researchers	.4728	–.1086	.1553	.4062	.1296	.5679
X_8	Participation in field days organized by researchers	.4183	–.1632	.3808	.0168	.2530	.5387
X_9	Attending package of practices workshops	.4018	–.2255	.4248	–.0641	.2630	.4687
X_{10}	Involving researchers in seminars	.3132	–.2602	.2377	.0937	.1574	.2588
X_{11}	Taking farmers to rice res. stations	.0736	–.3437	.1858	.0077	.1840	.2930
X_{12}	Taking help of researchers in laying out minikit trials	.1256	–.7760	.0694	.1195	.1329	.6827
X_{13}	Taking help of researchers in laying out demonstrations	–.0256	–.7731	.1019	.1936	.2057	.6933
X_{14}	Involving researchers in exhibitions	.1553	–.4309	.2325	.0293	.3208	.3969
X_{15}	Participation in exhibitions at res. stations	.3362	–.2372	.4036	–.0589	.3862	.5048
X_{16}	Participation in workshops organized by researchers	.3731	–.1872	.4755	–.1002	.3628	.5855
X_{17}	Participation in rice production technology training	.1213	–.0290	.6812	.1747	.0556	.5222
X_{18}	Participation in special training on rice at res. stations	.1048	–.1729	.6984	.2226	.2445	.5789
X_{19}	Involvement in campaigns organized by researchers	.2246	.3592	.3085	.1908	.3168	.4260
X_{20}	Giving suggestions to take up new area of res. on rice	.2963	–.0850	.2024	.1609	.5785	.4976
X_{21}	Co-operating with researchers in lab-to-land programme	.0665	–.0851	.0610	.1712	.6084	.4213
	Proportionate variance accounted by each factor	.1474	.0980	.1249	.0460	.0922	

Factor II

This factor was found to have contributed 9.80 per cent variability. Factor II had only two linkage activities with significant but negative loadings, X_{12}-taking help of researchers in laying out minikit trials (–.7760) and X_{13}-taking help of researchers in laying out demonstrations (–.7731). This factor could be designated as "help-seeking".

Though minikit trials and demonstrations are carried out in farmers' field by extension personnel, it is a joint venture of extension and research sub-systems. The trials and demonstration programmes should be developed jointly by the extension personnel and researchers and also analyse the results jointly. These provide a mechanism for closer and continuous working relationship between the two sub-systems.

Factor III

Factor III was found to have contributed 12.49 per cent variability being the highest after Factor I. This factor consisted of three linkage activities viz. X_{18} participation in special training on rice (.6984), X_{17}-participation in rice production technology training (.6812) and X_{16}-participation in workshops organised by researchers (.4755). All the three activities had significant and positive loadings. Considering the nature of linkage activities, Factor III could be conveniently named as "knowledge updating" which is a *sine quo non* for professionals engaged in the transfer of fast changing agricultural technologies.

Factor IV

Factor IV had only one linkage activity with factor loading above the limit of 0.45, this being X_2-participation with researchers in laying out farm trials (0.5033). The proportions contribution of this factor to the total variability was 4.60 per cent.

Factor V

Factor V had only two linkage activities having significant factor loadings viz. X_{21}-cooperating with researchers in lab-to-land programme (.6084) and X_{20}-giving suggestions to researchers on new

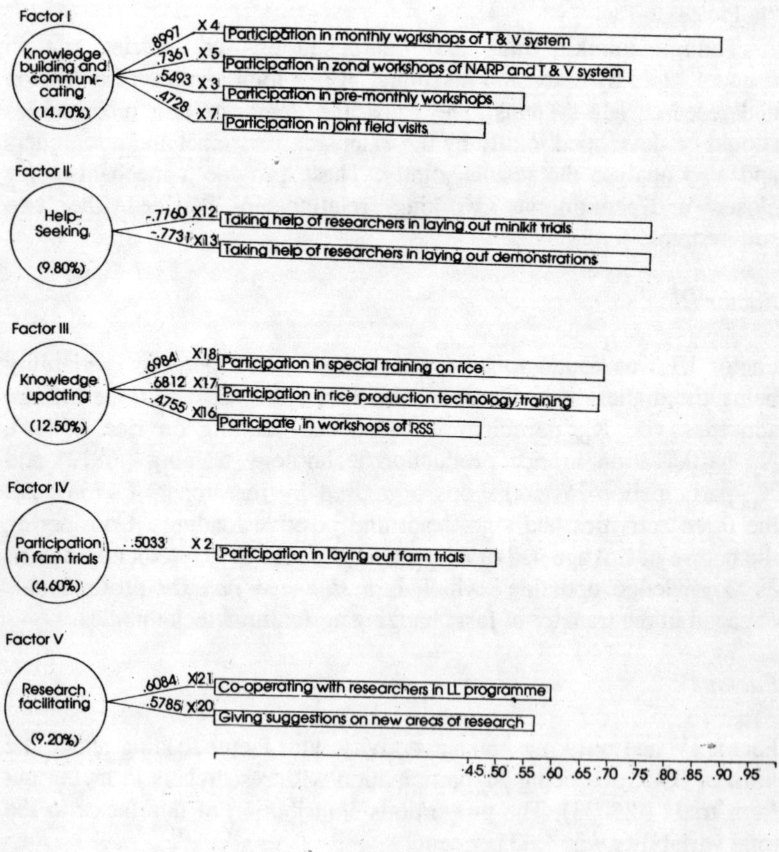

FIG. 4.7: Diagrammatic representation of principal-factor solutions for ESS-RSS linkage

areas of rice research to be taken up (.5785). The contribution of Factor V to the total variability was 9.20 per cent. This factor could be termed as "research facilitating".

4.3.5 *ESS-CSS Linkage*

Principal-factor solutions for ESS-CSS linkage are presented in Table 4.6. As a result of varimax rotation, six factors were extracted from the 19 linkage activities. The contribution of all the six factors together towards total variability was 46.40 per cent.

The nature and content of the individual factors are described below:

Factor I

The proportionate contribution of variance by Factor I was 9.69 per cent. This factor had only three linkage activities with significant loadings viz. X_{17}-keeping contact with local farmers' association (.5648), X_{15}-organising campaigns (.5559) and X_{19}-recognising innovative farmers (.4665). Considering the nature of individual activities, the common name traced out for Factor I was "liaising-motivating".

Factor II

The contribution of this factor to the total variability was 6.06 per cent. Factor II had only two linkage activities with significant loadings and they were X_1-visits to paddy fields of contact and other farmers (.6595) and X_2-visits to farm trials on rice (.6466). Both the activities were positively correlated to the factor and it could be rightly be called as "Supervising" factor.

Factor III

Factor III had also only two linkage activities having significant loadings: X_{11}-visiting farmers' fields on request (–.8069) and X_{12}-arranging for timely supply of material inputs (–.7484). Of these X_{11} was negatively correlated and X_{12} was positively correlated to the factor. The proportionate contribution of variability by Factor III was 8.03 per cent. The common name traced out for Factor III was "advising-enabling".

TABLE 4.6: Principal-factor solutions for extent of linkage of extension sub-system with client sub-system— factor patterns and their loadings

	Activities	Factor						Communality (h^2)
		I	II	III	IV	V	VI	
X_1	Visits to paddy fields of contact and other farmers	.2403	.6595	–.0983	–.1246	–.0360	–.1666	.5693
X_2	Visits to farm trials on rice	.0854	.6466	.0535	–.1292	–.0062	–.0442	.4713
X_3	Involvement in organizing farmers' days	.3530	.1977	.0575	–.1719	.0951	–.0299	.2798
X_4	Organizing farmers group discussions on rice	.1412	.2537	–.2082	–.5159	.0931	–.0246	.4042
X_5	Inviting farmers for seminars	.0468	.0271	–.0790	–.7267	.0008	.0465	.5546
X_6	Contribution of articles on rice in newspaper columns	.4088	.1325	–.1209	–.0666	.6398	.3628	.7448
X_7	Contribution of articles on rice in popular publications	.2812	–.0135	–.1604	–.0331	.7284	.3743	.7814
X_8	Providing information support to farmers through AIR	.0626	–.0144	–.0274	–.1342	.1459	.6492	.4816
X_9	Contributing articles to publications of farmers' association	.1747	.0247	–.0106	–.1817	.2187	.6513	.5228
X_{10}	Conducting demonstrations on rice	.4150	.1686	–.0109	–.0805	.4270	.1181	.2391
X_{11}	Visits to farmers' rice fields on request	.0496	.0843	–.8069	–.2819	–.1607	.1171	.7546
X_{12}	Arranging for timely supply of material inputs	.0776	.0205	–.7484	–.1691	–.0494	.1367	.6149
X_{13}	Arranging for timely supply of cash inputs	.2626	.1245	–.0863	–.2034	–.2648	.3644	.3363
X_{14}	Organizing exhibitions on rice for farmers	.0237	.0991	–.0674	–.0719	.4313	.0455	.2130
X_{15}	Organizing campaigns on rice	.5559	–.0609	–.0171	–.2154	–.0053	–.0533	.1995
X_{16}	Organizing training of rice production technology to farmers	.4363	.2097	.0089	.0046	.1103	.1430	.3436
X_{17}	Keeping contact with local farmers' association	.5648	.1074	.0485	–.1204	–.1560	.0778	.4255
X_{18}	Attending discussions organized by farmers	.2529	.0989	–.4133	–.5600	–.0147	.0411	.5609
X_{19}	Giving recognition to innovative farmers	.4665	.1772	–.1595	–.3790	–.0513	.1435	.5290
	Proportionate variance accounted by each factor	.0969	.0606	.0803	.0816	.0720	.0728	

Factor IV

This chapter contributed to 8.16 per cent variability to the hypothetical construct. Factor IV was dominated by three linkage activities viz. X_5-inviting farmers for seminars (–.7267), X_{18}-attending discussions organised by farmers (–.5600) and X_4-organising group discussions of farmers (–.5159). All the three variables were having fairly high loadings but were negatively correlated to the factor. This factor could be described as "interacting-enabling".

Factor V

The proportionate variability accounted for by Factor V was 7.20 per cent. This factor was dominated by two linkage activities with significant and positive loadings. They were X_7-contribution of articles in popular publications (.7284) and X_6-contribution of articles in newspaper columns (.6398). Both the activities were positively correlated to the Factor V. This factor could be termed as "Communicating".

Factor VI

Factor VI was also dominated by two linkage activities with significant positive loadings viz. X_9-contributing articles in publications of farmers' associations (.6513) and X_8-providing information support to farmers on rice through AIR (.6492). The proportionate contribution of variability of this factor was 7.20 per cent. This factor could be termed as "communicating", complimenting the activities identified under Factor V.

4.3.6 *ESS-ISS linkage*

Principal-factor solutions for ESS-ISS linkage are presented in Table 4.7. Five factors have been extracted from the 16 linkage activities of ESS with ISS and the five factors collectively contributed 51.00 per cent variability to the linkage of ESS with ISS.

The content and nature of individual factors are described below:

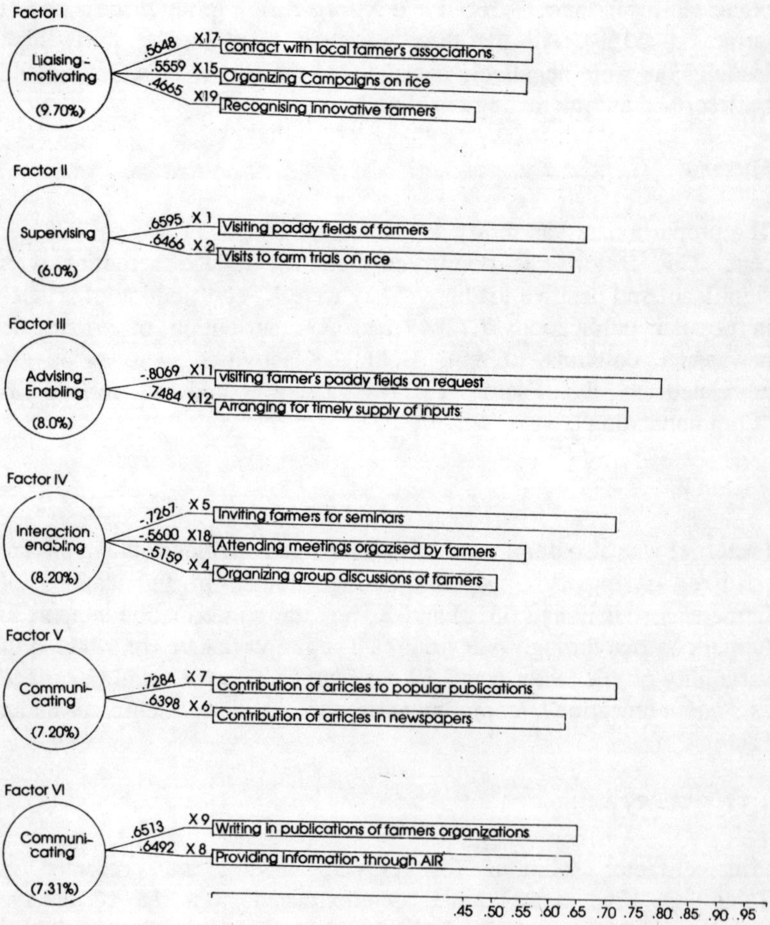

FIG. 4.8: Diagrammatic representation of principal-factor solutions for ESS-CSS linkage

TABLE 4.7: Principal-factor solutions for extent of linkage of extension sub-system with input sub-system—factor patterns and their loadings

	Activities	Factor					Communality (h^2)
		I	II	III	IV	V	
X_1	Arranging for supply of inputs during season	.6764	.4944	.4207	–.2752	–.0066	.9549
X_2	Making available credits to farmers in time	.3859	.2069	–.0746	–.2153	–.4905	.4844
X_3	Inviting input agencies for farmers' days	.5107	.0885	–.6146	–.2685	.0570	.7217
X_4	Involving input agencies in seminars	.4916	.2655	–.4238	–.0516	.0354	.4957
X_5	Involving input agencies in exhibitions	.4842	.0693	–.3205	.0374	.2484	.4051
X_6	Inviting input agencies on SMS training days	.0956	–.0124	.0476	–.0698	.4912	.2577
X_7	Checking the quality of fertilizers/pesticides	.5782	–.2719	.1027	–.3052	.1802	.5444
X_8	Taking steps to open sales points wherever required	.5394	.0051	.0203	–.3067	.3048	.4784
X_9	Keeping touch with input agencies for latest information	.4835	.1244	.2773	–.1451	.2639	.4168
X_{10}	Contributing articles to publication of input agencies	.0560	–.0092	.0390	.0690	.6253	.4005
X_{11}	Participation in seminars organized by input agencies	.5699	–.0905	.0238	.4699	.2450	.6145
X_{12}	Participation in campaigns organized by input agencies	.5695	.0348	–.0227	.5076	.2847	.6649
X_{13}	Participation in exhibitions organized by input agencies	.6011	–.0422	.0344	.3887	.3147	.6145
X_{14}	Organising training for input agencies	.4374	–.4937	.1581	.0657	.1716	.4939
X_{15}	Providing consultancy service to input agencies	.4846	–.3371	.1387	–.0232	.2933	.4565
X_{16}	Conveying complaints of farmers to input agencies	.4588	–.0778	.0522	–.0570	.0550	.2255
	Proportionate variance accounted by each factor	.2415	.0522	.0615	.0659	.0933	

Factor I

As it could be seen from Table 4.7 Factor I was influenced by 12 linkage activities with significant loadings well above the limit of .45. The proportionate contribution of Factor I towards total variability was 24.15 per cent. Taking the linkage activities individually, activity X_1-arranging for supply of inputs during season had the highest factor loading (.6764). This was followed by X_{13}-participation in exhibitions organised by input agencies (.6011), X_7-checking the quality of fertilizers/pesticides (.5782), X_{11}-participation in seminars organised by input agencies (.5699), X_{12}-participation in campaigns organised by input agencies (.5695), X_8-taking steps to open sales points wherever required (.5394), X_3-inviting input agencies for farmers' days (.5107), X_4-involving input agencies in seminars (.4916), X_{15}-providing consultancy services to input agencies (.4846), X_5-involving input agencies in exhibitions organised by ESS (.4842), X_9-keeping touch with input agencies for latest information (.4835) and X_{16}-conveying complaints of farmers to input agencies (.4588) in the descending order of factor loadings.

All the 12 linkage activities were significantly and positively correlated to the Factor I. From the point of view of the nature of linkage activities, Factor I could be designated as "Interacting-regulating".

Factor II

Factor II had only two linkage activities with significant loadings – X_1-arranging for supply for inputs during season (.4944) and X_{14}-organising training for input agencies (–.4937). Of the two linkage activities, X_1 was positively loaded whereas, X_{14} was negatively loaded. Factor II contributed 5.22 per cent variability towards the linkage of ESS with ISS. This factor could be designated as "liaising-training".

The linkage activity X_1 dominated under Factor I was also having the highest factor loading among the 12 activities. Presence of the same linkage activity under two factors consecutively with highest factor loadings indicate the exigency of the efficient distribution system for inputs like seeds, fertilizers, pesticides, credit etc. on the part of the extension sub-system. The importance is two-fold. Firstly it helps to strengthen the linkage between ESS and ISS and secondly the TOT process could take better momentum.

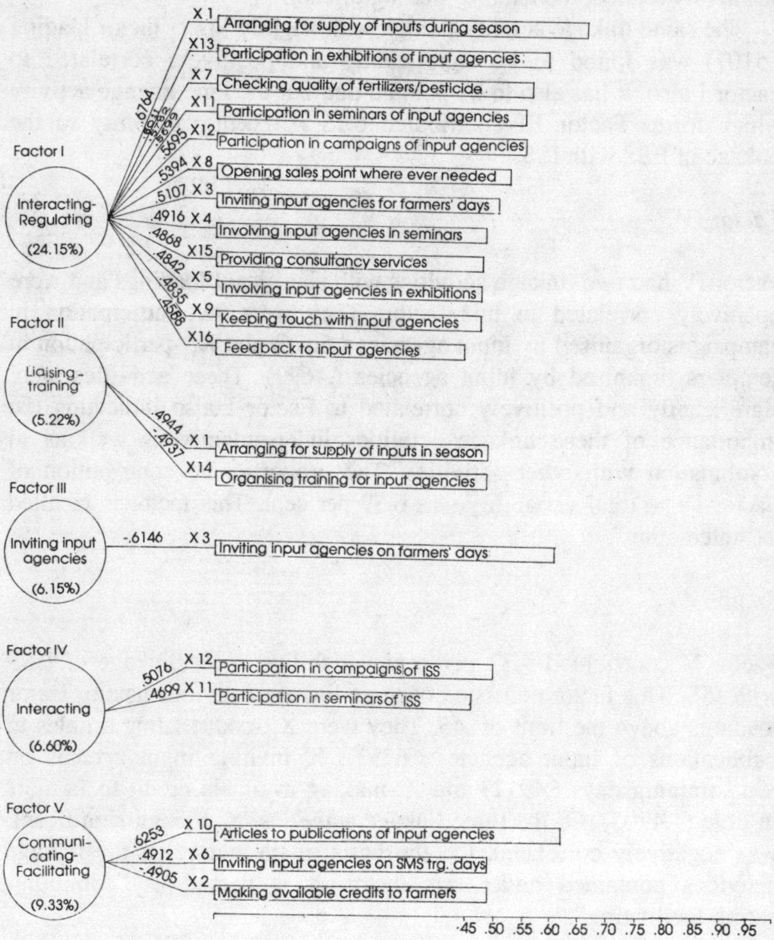

FIG. 4.9: **Diagrammatic representation of principal-factor solutions for ESS-ISS linkage**

Factor III

In Factor III, there was only one linkage activity with significant loading, X_3-inviting input agencies on farmers' days (–.6146). Though negatively related, the loading was significant.

The same linkage activity though with slightly lower factor loading (.5107) was found to be significantly and positively correlated to Factor I also. It has also to be pointed out that the lone linkage activity which forms Factor III contributed 6.15 per cent variability to the linkage of ESS with ISS.

Factor IV

Factor IV had two linkage activities with significant loadings and were positively correlated to this factor. They were X_{12}-participation in campaigns organised by input agencies (.5076) and X_{11}-participation in seminars organised by input agencies (.4699). These activities were significantly and positively correlated to Factor I also indicating the importance of these linkage activities independently as well as in combination with other activities. The proportionate contribution of Factor IV to total variability was 6.59 per cent. This factor is labelled as "interacting".

Factor V

Factor V contributed 9.33 per cent variability to the linkage of ESS with ISS. This factor consisted of three linkage activities having factor loadings above the limit of .45. They were X_{10}-contributing articles to publications of input agencies (.6253), X_6-inviting input agencis on SMS training days (.4912) and X_2-making available credit to farmers in time (–.4905). Of the three linkage activities, X_2 though significant, was negatively correlated. On the basis of the nature of the linkage activities contained under this factor, it is named as "communicating-facilitating".

4.3.7 *CSS-RSS Linkage*

Principal-factor solutions for CSS-RSS linkage are presented in Table 4.8.

TABLE 4.8: Principal-factor solutions for extent of linkage of client sub-system with research sub-system—factor patterns and their loadings

	Activities	Factor				Communality (h^2)
		I	II	III	IV	
X_1	Visits to rice research stations	.4782	.1169	.4056	.3212	.5100
X_2	Participation in training organized by rice research stations	.5364	–.0246	.3264	.2078	.4381
X_3	Contacting researchers for specific recommendations on rice	.6714	–.0882	.0322	.0765	.4654
X_4	Participation in kisan melas organised by rice research stations	.7087	–.4323	.0896	–.0828	.7041
X_5	Participation in exhibitions organised by rice research stations	.6307	–.1618	.2252	.0900	.4828
X_6	Participation in farmers day organized by rice research stations	.6777	–.2683	.3059	–.0802	.6313
X_7	Obtaining seeds from rice research stations	.5387	–.2128	.0665	.3111	.4366
X_8	Co-operating with researchers in the conduct of trials	.3965	–.2552	.4215	.0221	.4006
X_9	Making correspondence with rice research stations	.2522	–.1297	.0877	.5144	.3527
X_{10}	Reading the articles on rice in newspapers	.1311	–.5990	.1482	–.2063	.4406
X_{11}	Reading the articles on rice in popular publications	.1918	–.5625	.1730	.1844	–.4172
X_{12}	Listening to the talk of researchers through AIR	–.1215	–.5765	.1600	–.0547	.3757
X_{13}	Giving land for conducting trials by researchers	.3335	–.3963	.2686	.0746	.3460
X_{14}	Visits to the trials in other farmers' fields	.1416	–.3241	.4726	.1295	.3652
X_{15}	Seeking the help of researchers to solve problem on rice	.6068	–.2351	–.0783	.2446	.4894
X_{16}	Getting benefits (inputs) through lab-to-land programme	.1724	–.6076	–.2952	.1230	.5012
X_{17}	Getting benefits (inputs) through village adoption programme	.1552	–.6440	–.0118	.3209	.5419
	Proportionate variance accounted by each factor	.2035	.1496	.0636	.0480	

As a result of varimax rotation four factors have been extracted from 17 linkage activities. The contribution of four factors towards total variability was 46.47 per cent. The content and nature of individual factors are given below:

Factor I

As many as eight linkage activities were found to have significant loadings well above the limit of .45 and the proportionate contribution of variability was 20.35 per cent. These activities in the descending order of factor loadings were X_4-participation in kisan melas organised by rice research stations (.7087), X_6-participation in farmers' day organised at research stations (.6777), X_3-contacting researchers for specific recommendations on rice (.6714), X_5-participation in exhibition organised at research stations (.6307), X_{15}-seeking help of researchers in solving problems on rice (.6068), X_7-obtaining seeds from rice research stations (.5387), X_2-participation in training organised by research station (.5364) and X_1-visits to rice research stations (.4782). All the eight linkage activities were positively related to the Factor I and this factor could be reasonably called as "innovation seeking" since farmers indulge in these activities only if they have the quest for new knowledge.

Factor II

Factor II contributed 14.96 per cent variability to the CSS-RSS linkage. It was found that five linkage activities were having significant factor loading under Factor II. They included X_{17}-getting inputs through village adoption programme (–.6440), X_{16}-getting benefits through lab-to-land programme (–.6076), X_{10}-reading articles on rice in newspapers (–.5990), X_{12}-listening to the talk of researchers through AIR (–.5765) and X_{11}-reading the articles on rice in popular publications (–.5625). One notable feature of Factor II was that though all the five activities were significantly related, they were negatively correlated to the factor. Factor II could be designated as "Incentive seeking-Information seeking" due to the nature of the activities contained therein.

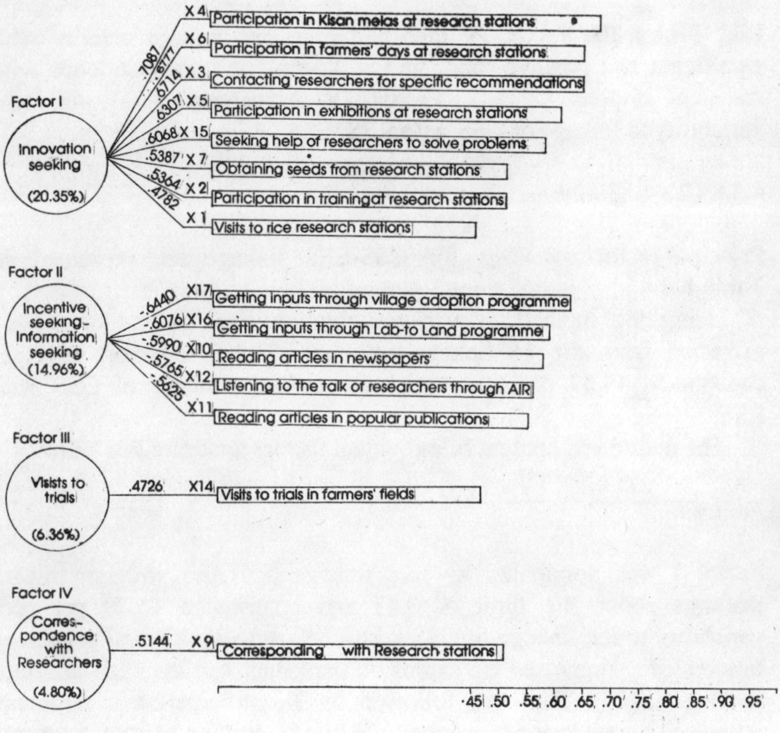

FIG. 4.10: Diagrammatic representation of principal-factor solutions for CSS-RSS linkage

Factor III

This factor had only one linkage activity with significant loading viz. X_{14}-visits to the trials in other farmers' fields (.4726) and the proportionate contribution to total variance was 6.36 per cent.

Factor IV

Like Factor III, Factor IV also had only one linkage activity with significant and positive loadings i.e. X_9-making correspondence with research stations (.5144). Factor IV contributed 4.80 per cent variability to linkage of CSS with RSS.

4.3.8 *CSS-ESS Linkage*

Principal-factor solutions for CSS-ESS linkage are presented in Table 4.9.

Using the method of varimax rotation five factors have been extracted from the 16 linkage activities. The five factors together contributed 49.57 per cent variability towards linkage of CSS with ESS.

The nature and content of individual factors are described below:

Factor I

Factor I was dominated by five linkage activities with significant loadings above the limit of 0.45 and contributed 15.70 per cent variability to the linkage of CSS with ESS. Activity X_4-participation in farmers' day organised by extension personnel had the highest factor loading (.8424). This was followed by X_5-participation in seminars organised by extension personnel (.7939), X_7-seeking help of extension workers in solving problems on rice (.5009), X_3-participation in demonstrations laid out by extension personnel (.4772) and X_6-participation in exhibitions organised by extension personnel (.4647). Taking the nature of the linkage activities into consideration, Factor I could be described as "knowledge seeking-problem solving".

TABLE 4.9: Principal-factor solutions for extent of linkage of client sub-system with extension sub-system—factor patterns and their loadings

	Activities	Factor					Commun-ality (h^2)
		I	II	III	IV	V	
X_1	Participation in training organized by extension workers	.4178	–.1254	.0047	.6763	–.1273	.6640
X_2	Co-operating in farm trials conducted by extension personnel	.0715	–.2060	–.1740	.6326	.0882	.4853
X_3	Participation in demonstration laid out by extension personnel	.4772	–.2183	.0541	.4263	.0481	.4604
X_4	Participation in farmers' day organized by extension personnel	.8424	–.1055	.0519	.2485	.0487	.7876
X_5	Participation in seminars organized by extension personnel	.7939	–.2409	–.1521	.1774	–.0388	.7436
X_6	Participation in exhibitions organized by extension personnel	.4647	.2964	–.0085	.2414	–.1193	.4457
X_7	Seeking help of extn. workers for solving problems on rice	.5009	.1219	–.3001	.0504	–.1618	.3846
X_8	Participation in discussions organized by extension personnel	.3297	–.0200	–.2796	.6340	–.2366	.6452
X_9	Reading articles in newspapers written by extension personnel	–.0170	–.6954	.0830	.1300	.0071	.5239
X_{10}	Reading articles in popular publications	.1738	–.5746	–.0460	.0892	–.1844	.4044
X_{11}	Visit to the agricultural extension units	.0317	–.1943	–.4161	.1453	.1832	.2675
X_{12}	Listening to the talk given by extension personnel through AIR	.2799	–.5783	–.0678	.0753	–.2180	.4705
X_{13}	Contacting the extension personnel for inputs	.1039	–.1042	–.5030	.1167	–.1414	.3084
X_{14}	Sharing problems/experiences with extension workers	.1945	–.0914	–.6107	.0543	–.1232	.4873
X_{15}	Corresponding with extension workers	.1131	–.2204	–.0674	.1034	–.5602	.3904
X_{16}	Inviting extension personnel in discussions	.1337	–.1447	–.1021	.3798	–.3731	.3925
	Proportionate variance accounted by each factor	.1570	.1012	.0659	.1146	.0570	

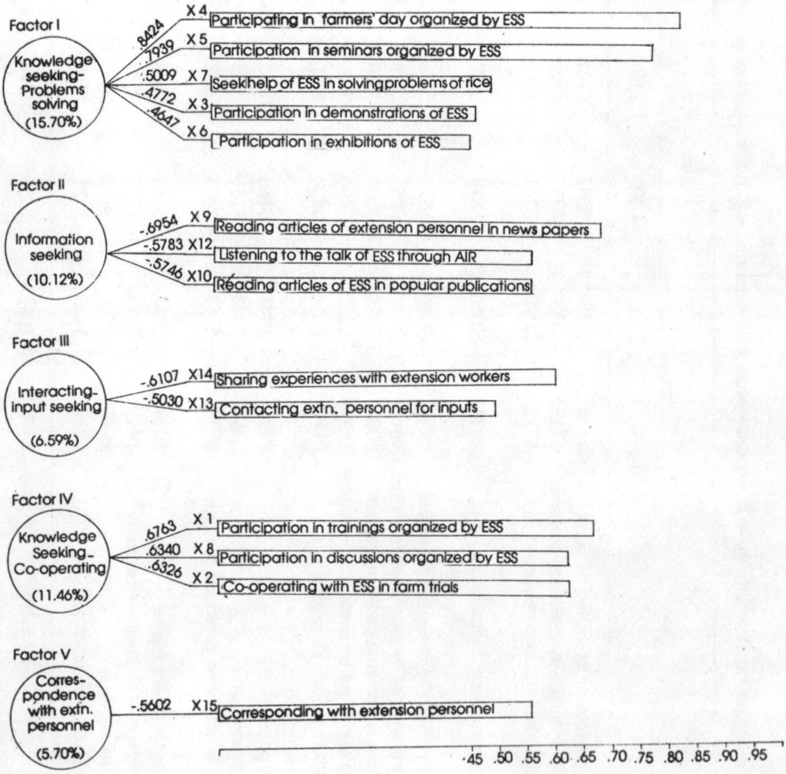

FIG. 4.11: **Diagrammatic representation of principal-factor solutions for CSS-ESS linkage**

Factor II

Factor II was influenced by three linkage activities X_9, X_{12} and X_{10}, reading articles in newspapers written by extension personnel (–.6954), listening to the talk of extension personnel through AIR (–.5783) and reading articles in popular publications (–.5746) respectively. All the activities though significantly loaded, were negatively related to Factor II and the proportionate contribution towards linkage was 10.12 per cent. This factor could be termed as "information seeking".

Factor III

This factor had only two linkage activities with significant loadings. They were X_{14}-sharing problems/experiences with extension workers (–.6107) and X_{13}-contacting the extension personnel for inputs (–.5030). Both the activities were negatively related but Factor III as a whole contributed 6.59 per cent variability to linkage. This factor could be called as "Interacting-Input seeking".

Factor IV

Factor IV was mainly influenced by three linkage activities X_1-participation in trainings organised by extension workers (.6763), X_8-participation in discussions organised by extension personnel (.6340) and X_2-cooperating in farm trials conducted by extension personnel (.6326). All the three activities had significant positive factor loadings and the proportionate contribution of this factor to total variability was 11.46 per cent. Factor IV could be reasonably termed as "knowledge seeking-co-operating".

Factor V

The only linkage activity with significant loading in this factor was X_{15}-corresponding with extension workers (–.5602). Though it was negatively related to Factor V, this factor contributed 5.70 per cent variance to linkage of CSS with ESS.

4.3.9 *CSS-ISS Linkage*

Principal-factor solutions for CSS-ISS linkage are shown in Table 4.10.

TABLE 4.10: Principal-factor solutions for extent of linkage of client sub-system with input sub-system—factor patterns and their loadings

	Activities	Factor			Communality
		I	II	III	(h^2)
X_1	Collecting technical information from input agencies	.4218	.4482	–.0061	.3789
X_2	Participation in seminars organized by input agencies	.8106	.1799	–.1592	.7149
X_3	Participation in discussions of input agencies	.8935	.1592	–.1744	.8541
X_4	Visiting exhibitions organized by input agencies	.7188	.1872	–.1328	.5694
X_5	Co-operating with input agencies in demonstrations	.5384	.1518	–.2776	.3900
X_6	Visiting the result demonstration of input agencies	.4761	.0087	–.4187	.4014
X_7	Availing credit facilities from credit agencies	.1149	.0449	–.5064	.2717
X_8	Knowing about the services available with input agencies	.2859	.5232	–.3964	.5127
X_9	Taking help of input agencies in preparing production plan	.2911	.5720	.0166	.4122
X_{10}	Reading the publication of input agencies	.3666	.4788	–.0603	.3051
X_{11}	Feeding back opinion about quality of inputs	.2372	.4068	–.0806	.2295
X_{12}	Feeding back to the input agencies about their performances	.2488	.4866	–.4582	.5086
	Proportionate variance accounted by each factor	.2582	.1241	.0800	

As it could be seen from Table 4.10 three factors have been extracted from the 12 linkage activities. All the three factors jointly contributed 46.23 per cent variability to linkage of CSS with ISS.

The descriptions of the individual factors are given below:

Factor I

The proportionate variance accounted by Factor I was 25.82 per cent. This factor had five linkage activities with loadings well above the limit of 0.45. Among these, X_3-participation in discussion meetings of input agencies had a very high factor loading of .8935. This was followed by linkage activities X_2-participation in seminars organised by input agencies (.8106), X_4-visits to exhibitions organised by input agencies (.7188), X_5-cooperating with input agencies in demonstrations (.5384) and X_6-visiting the result demonstrations of input agencies (.4761). The common name traced out for this factor is "knowledge seeking-cooperating".

Factor II

Factor II had four linkage activities with significant loadings viz. X_9-taking help of input agencies in preparing production plan (.5720), X_8-knowing about the services available with input agencies (.5232), X_2-feeding back to the input agencies about their performances (.4866) and X_{10}-reading the publications of input agencies (.4788). The proportionate variance accounted by Factor II was 12.41 per cent and this factor could be called as "help seeking-communicating".

Factor III

This factor contributed 8.00 per cent variability to the linkage and consisted of two linkage activities with significant negative projections. They were X_7-availing credit facilities from credit agencies (–.5064) and X_{12}-feeding back to input agencies about their performances (–.4582). Factor III could be termed as "help seeking-communicating" similar to that in the case of Factor II.

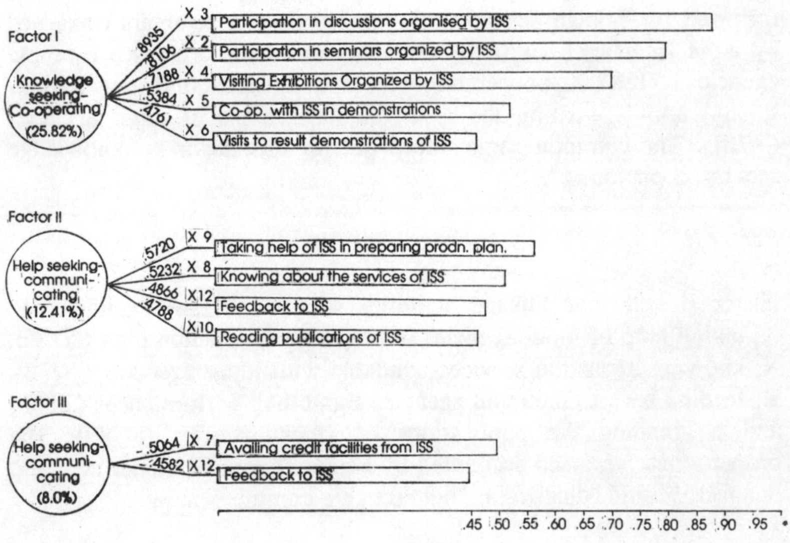

FIG. 4.12: Diagrammatic representation of principal-factor solutions for CSS-ISS linkage

4.3.10 *ISS-RSS Linkage*

Principal-factor solutions for ISS-RSS linkage are presented in Table 4.11. As it is evident from the table as a result of varimax rotation four factors have been extracted from 13 linkage activities of ISS with RSS. All the four factors jointly contributed 64.80 per cent variability towards the extent of linkage of ISS with RSS. The nature and content of individual factors are given below:

Factor I

The first factor had the highest proportionate variability of 26.43 per cent. This factor had as many as nine linkage activities with significant factor loadings above the limit of 0.45. X_7-mailing publications to rice research stations possessed the highest loading of .7400. The loadings for the other linkage activities in the descending order were X_{12}-extending financial help to research stations for melas/field days (.7109), X_3-collecting information on new rice varieties released by the KAU (.6920), X_6-participation in package of practices workshops (.6441), X_4-participation in field days organised at rice research stations (.5829), X_{13}-contributing articles to research publications (.5623), X_2-getting new fertilizers/pesticides tested at research stations (.5258), X_1-keeping contact with scientists of rice research stations (.4744) and X_8-inviting researchers in melas organised by input agencies (.4600). Taking the nature and content of the first factor, it could be reasonably named as "facilitating-validating-interacting".

Factor II

Factor II was influenced mainly by three linkage activities with significant but negative loadings. Of these, activity X_9-feeding back the field experiences to researchers had a very high loading (–.9601). Activity X_{10}-taking advice of researchers in laying out demonstrations had the next high loading of –.6586 and activity X_8-inviting researchers for melas organised by input agencies carried a loading of –.5104. The proportionate variance accounted by Factor II to total variability was 13.40 per cent. This factor could be termed as "Facilitating — knowledge building" on the basis of the general nature of the activities coming under the factor.

TABLE 4.11: Principal-factor solutions for extent of linkage of input sub-system with research sub-system—factor patterns and their loadings

	Activities	Factor				Communality (h^2)
		I	II	III	IV	
X_1	Keeping contact with scientists of rice research stations	.4744	–.1140	.3850	.3366	.4996
X_2	Getting new pesticides/fertilizers tested at research stations	.5258	–.0486	.0624	.5407	.5751
X_3	Collecting information on new rice varieties released by KAU	.6920	–.3747	–.0222	–.0838	.6238
X_4	Participation in field days organised at rice research stations	.5829	–.0736	.6218	.0466	.7368
X_5	Participation in exhibitions at rice research stations	.1387	–.1843	.8471	.2018	.8116
X_6	Participation in package of practices workshops	.6441	.0785	.0650	.3428	.5428
X_7	Mailing publications to rice research stations	.7400	–.1990	.0918	.2659	.6663
X_8	Inviting researchers in meals organized by input agencies	.4600	–.5104	.4019	.1181	.6478
X_9	Feeding back to researchers about field experiences	–.0100	–.9601	.2859	.0445	.8237
X_{10}	Taking advice of researchers in laying out demonstrations	.2545	–.6586	–.1037	.4811	.7408
X_{11}	Giving fertilizers/pesticides to research stations for trials	.1823	–.0775	.1461	.7524	.6268
X_{12}	Extending financial help to research stations for melas/fields days	.7109	–.1709	.2592	.2868	.6841
X_{13}	Contributing articles to research publications	.5623	–.1782	.2856	.1255	.4453
	Proportionate variance accounted by each factor	.2643	.1340	.1303	.1194	

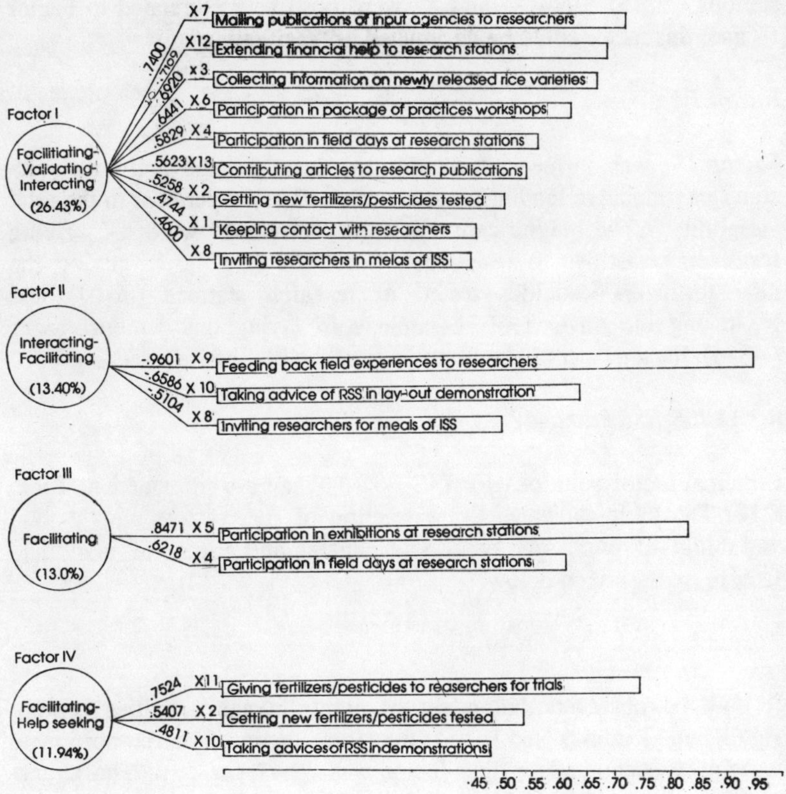

FIG. 4.13: Diagrammatic representation of principal-factor solutions for ISS-RSS linkage.

Factor III

Factor III contributed 13.03 per cent variability to the extent of linkage of ISS with RSS. It consisted of two linkage activities with significant loadings viz. X_5-participation in exhibitions at rice research stations (.8471) and X_4-participation in field days organized at rice research stations (.6218). Both X_5 and X_4 were positively correlated to Factor III and, this factor could be designated as "facilitating".

Factor IV

Factor IV was influenced mainly by three linkage activities with significant positive loadings and contributed 11.94 per cent to the total variability. The significant linkage activities were X_{11}-giving fertilizers/pesticides to researchers for trials (.7524), X_2-getting the new fertilizers/pesticides tested at research stations (.5407) and X_{10}-taking the advices of researchers in laying out demonstrations (.4811). Factor IV could be described as "facilitating-help seeking".

4.3.11 *ISS-ESS Linkage*

Principal-factor solutions for ISS-ESS linkage are presented in Table 4.12. The table indicates the extraction of five factors with a total variability of 64.82 per cent. The content and nature of individual factors are described below:

Factor I

It may be observed that a set of eight linkage activities bearing significant loadings explained the first factor. The proportionate variability accounted by this factor was 22.47 per cent. The largest weights appear for linkage activity X_{18}-cooperating the KAU in national demonstrations (.8336). Next to this were X_{14}-cooperating with the KAU in lab-to-land programme (.7436), X_{15}-cooperating with KAU in village adoption programme (.7217), X_6-mailing literature to extension personnel (.6778), X_4-participation in exhibitions organised by extension agencies (.5548), X_{17}-financing extension agencies for seminars (.5458), X_9-taking help of extension agencies for tackling field problems (.5165) and X_{13}-attending fortnightly trainings organised by SDAOs (.5001). Factor I could be appropriately be called as "cooperating-help seeking —interacting" due to the basic nature of the activities included under this factor.

Factor II

Factor II had the highest proportionate variability (24.44%). Factor II was influenced by eight linkage activities viz. X_3- participation in seminars organised by extension agencies (.8928), X_2-participation in melas organised by extension agencies (.8090), X_7 -inviting extension personnel for seminars (.7795), X_8-inviting extension personnel in campaigns (.6676), X_{10}-keeping constant contacts with extension personnel (.6512), X_1-participation in campaigns organised by extension agencies (.6440), X_4-participation in exhibitions organised by extension agencies (.6005) and X_{17}-financing extension agencies for seminars/workshops (.4823). Linkage activity X_{17} was also one of the significant components of Factor I with a higher factor loading. Factor II could be conveniently called as "interacting-help seeking-facilitating".

Factor III

Factor III consisted of only one linkage activity X_{12}-organising special training for students of agriculture with a significant positive loading of .6886. The proportionate variance accounted by this factor was 6.97 per cent. This activity is being performed only by the fertiliser dealers.

Factor IV

Factor IV had two linkage activities with significant factor loadings viz. X_{11}-attending dealers training organised by SDAOs (.6007) and X_1-participation in campaigns organised by extension agencies (.4694). Its contribution to total variability was 5.68 per cent. This factor could logically be termed as "Interacting".

Factor V

Factor V contributed 5.26 per cent variability towards linkage of ISS with ESS. The lone linkage activity in this factor with significant loadings (.6813) was X_{16}-co-operating with extension personnel in farm trials.

TABLE 4.12: Principal factor solutions for extent of linkage of input sub-system with extension sub-system—factor patterns and their loadings

	Activities	Factor					Communa-lity (h^2)
		I	II	III	IV	V	
X_1	Participation in campaigns organised by extension agencies	.1620	.6440	.2932	.4694	.1619	.7677
X_2	Participation in melas organised by extension agencies	.0720	.8090	.2419	.3372	.1311	.8492
X_3	Participation in seminars organised by extension agencies	.0198	.8928	.1345	.2265	.1510	.8897
X_4	Participation in exhibitions organised by extension agencies	.5548	.6005	.2475	.0676	.0167	.7346
X_5	Keeping the extension agencies informed about latest products	.4184	.2560	.1640	.0137	.2328	.3219
X_6	Mailing literature to extension personnel	.6778	.2110	.2648	–.4186	.1360	.5944
X_7	Inviting extension personnel for seminars	.2202	.7795	–.2366	–.2436	–.1689	.8001
X_8	Inviting extension personnel for campaigns	.1656	.6676	.1384	–.3704	–.1106	.6417
X_9	Taking help of extension agencies for tackling field problems	.5165	.2475	–.0932	.0144	.0773	.3430
X_{10}	Keeping constant contacts with extension personnel	.3670	.6512	.3630	.0619	–.0385	.6959
X_{11}	Attending dealers training organised by SDAOs	.4150	.2294	.1950	.6007	.0866	.6313
X_{12}	Organising special training for students of agriculture	.0756	.1354	.6886	–.0090	.1541	.5221
X_{13}	Attending fortnightly training organised by SDAOs	.5001	.2503	–.0697	.1320	.1729	.3650
X_{14}	Co-operating with KAU in lab-to-land programme	.7436	.1194	.2560	.2019	.2415	.7320
X_{15}	Co-operating with KAU in village adoption programme	.7217	.1768	.0778	.0779	.4186	.7395
X_{16}	Co-operating with extension personnel in farm trials	.3223	3101	–.0917	–.0578	.6813	.6760
X_{17}	Financing extension agencies for seminars/workshops	.5458	.4823	–.3496	.1953	–.0282	.6910
X_{18}	Co-operating with KAU in national demonstrations	.8336	.1255	–.0010	.0226	.0343	.0119
	Proportionate variance accounted by each factor	.2247	.2444	.0697	.0568	.0526	

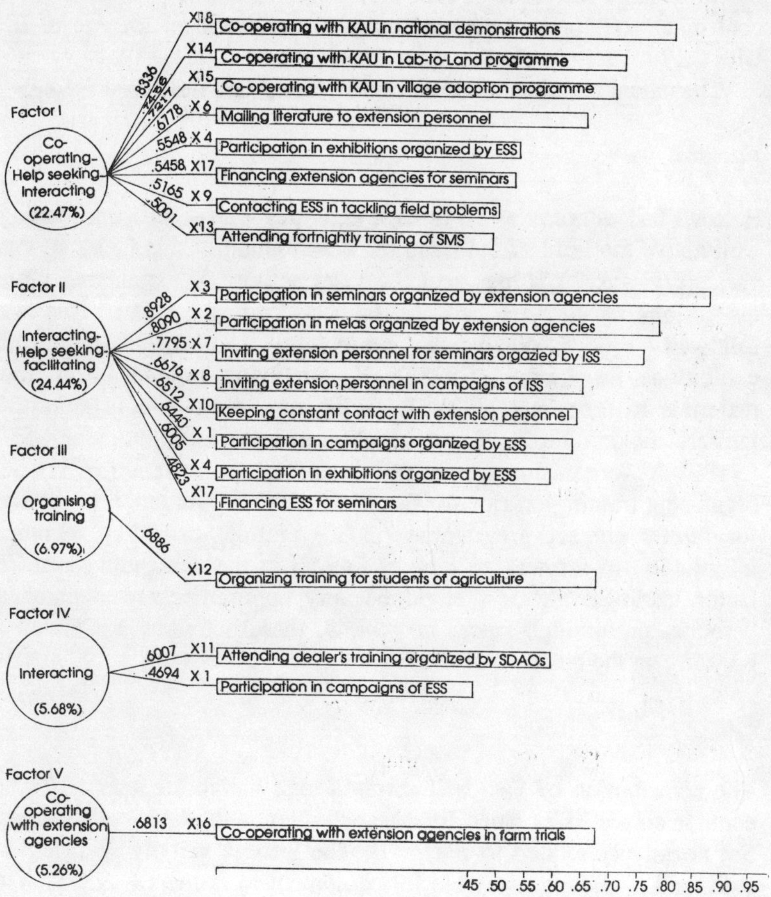

FIG. 4.14: Diagrammatic representation of principal-factor solutions for ISS-ESS linkage

4.3.12 *ISS-CSS linkage*

Principal-factor solutions for ISS-CSS linkage are presented in Table 4.13. As it is evident from the table, six factors emerged from the 19 linkage activities of ISS with CSS. All the six factors together contributed 67.20 per cent variability to the extent of linkage of ISS with CSS.

The nature and content of individual factors are described below:

Factor I

Factor I had as many as 10 linkage activities with significant loadings well above the limit of 0.45 and the contribution of this factor to total variability was 29.71 per cent. Linkage activity X_8-organising group discussions of farmers had the highest loading (.8429). This was followed by X_6-organising campaigns (.8394), X_9-organising exhibitions for farmers (.8287), X_{10}-supplying printed information materials to farmers (.7892), X_2-laying out result demonstrations in farmers' fields (.7362), X_7-participation in input fortnight celebrations (.7187), X_{19}-organising special farmer contact programme (.6731), X_5-organising training on rice for farmers (.6286), X_4-adoption of villages for farmer contact programme (.5080) and X_{14}-providing technical information to farmers on request (.4922) in the descending order of factor loadings. Factor I could be very appropriately designated as "product promoting" since, in general, these activities are aimed at boosting up the business of the ISS.

Factor II

The contribution of Factor II towards total variability was 10.74 per cent. It consisted of three linkage activities with significant loadings but negatively related to Factor II. The linkage activity with highest loading X_{17}-making available PP equipment to farmers (–.9303). This was followed by X_{18}-helping farmers in marketing rice (–.5557) and X_{12}-helping farmers in availing credit facilities (–.4928). Factor II could be termed as, "facilitating".

TABLE 4.13: Principal factor solutions for extent of linkage of input sub-system with client sub-system—factor patterns and their loadings

	Activities	Factor						Communality (h^2)
		I	II	III	IV	V	VI	
X_1	Timely supply of inputs to farmers	.0042	–.2230	.6300	.0566	.1454	–.0100	.4711
X_2	Laying out result demonstrations in farmers fields	.7362	–.1476	–.0075	.1506	.2226	.1074	.6477
X_3	Opening sales point at required places	.4444	–.4464	.3156	.0200	–.0607	–.1627	.5269
X_4	Adoption of villages for farmer contact programme	.5080	–.1235	.4421	–.1406	.2565	–.1558	.5787
X_5	Organising training on rice for farmers	.6281	–.2171	.0562	–.2026	–.1852	–.0079	.5209
X_6	Organising campaigns	.8394	–.0253	.1487	–.3332	.0684	.0870	.8506
X_7	Participation in input fortnight celebrations	.7187	–.0049	.3341	–.2382	–.1123	.1315	.7148
X_8	Organising group discussions of farmers	.8429	–.0567	.1208	–.1274	.0542	.2943	.8340
X_9	Organising exhibitions for farmers	.8287	–.1284	–.2200	.1221	.2434	.0728	.8310
X_{10}	Supplying printed information materials to farmers	.7892	–.1390	.0022	.2021	.0267	–.0691	.6885
X_{11}	Arranging film shows on rice production technology	.1796	–.1845	–.2790	–.4160	.2833	.3479	.5185
X_{12}	Helping farmers in availing credit facilities	.0147	–.4928	.5624	.0817	.2217	.0318	.6163
X_{13}	Publication of journals/newsletters for farmers	.3975	–.1647	–.0331	–.3373	.0132	.3750	.4408
X_{14}	Providing technical information to farmers on request	.4922	–.1768	.3085	.1427	–.3423	.6323	.9060
X_{15}	Receiving complaints of farmers properly	.3246	.0685	–.0039	.6124	.1990	.4762	.7515
X_{16}	Extending soil testing service to farmers	.0940	–.3338	.1040	–.1136	.8325	.1066	.8484
X_{17}	Making available PP equipments to farmers	.0695	–.9303	.0568	.1236	.0844	–.0335	.8970
X_{18}	Helping farmers in marketing rice	.0785	–.5557	–.0340	–.2620	.0174	.0815	.3917
X_{19}	Organising special farmer contact programme	.6731	–.2064	–.2914	.1970	.2846	.1876	.7356
	Proportionate variance accounted by each factor	.2971	.1074	.0780	.0606	.0696	.0593	

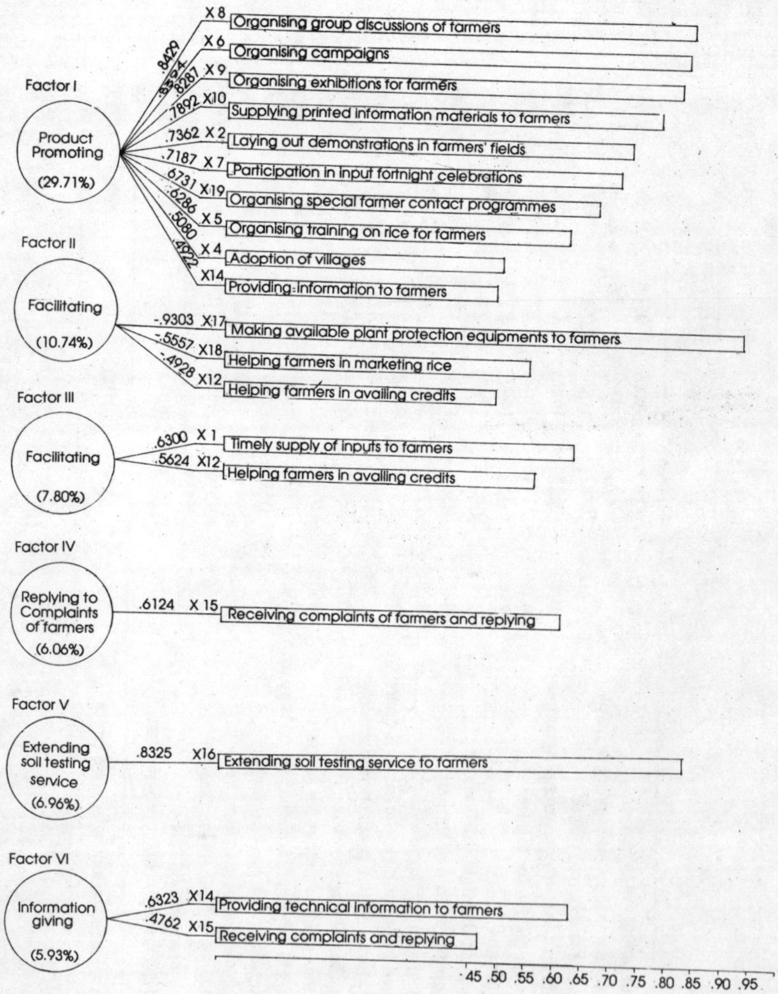

FIG. 4.15: Diagrammatic representation of principal-factor solutions for ISS-CSS linkage

Factor III

The proportionate variability accounted by Factor III was 7.80 per cent. It consisted of two linkage activities with significant loadings viz. X_1- timely supply of inputs to farmers (.6300) and X_{12}-helping farmers in availing credit facilities (.5624). Of these, X_{12} was found to be a contributing activity of Factor II also with slightly lesser and negative loading. Factor III could be called as "facilitating" as in the case of Factor II.

Factor IV

This factor had only one linkage activity with significant loading viz. X_{15}-receiving complaints from farmers and replying properly (.6124). This lone linkage activity contributed 6.06 per cent variability in the linkage of ISS with CSS.

Factor V

Factor V also had only one linkage activity X_{16}-extending soil testing service to farmers with a very high factor loading of .8325 and it contributed 6.96 per cent to the variability in the linkage of ISS with CSS.

Factor VI

Factor VI had two linkage activities with significant loadings. They were X_{14}-providing technical information to farmers on request (.6323) and X_{15}-receiving the complaints from farmers and replying properly (.4762). Of these, X_{15} was a contributing activity of Factor VI with slightly higher loading and X_{14} was a contributing activity to Factor I with slightly lesser factor loading. The fact that these two activities emerge as significant in more than one factor shows that these two activities have significant linking functions between the ISS and CSS. Factor VI could be described as "information giving". The proportionate contribution of this factor to total variability was 5.93 per cent.

4.4 Importance of linkage activities between and among research, extension, client and input sub-systems as perceived by the respondents – sub-systemwise analysis

For the interpretation of the results of the study on perception of the respondents of the different sub-systems about the importance of linkage activities performed by them and also to test the hypotheses set for the study, a weighted average score for each respondent was worked out and ranked. The sum of ranks secured by each sub-system for the perception of the importance of linkage activities performed with every other sub-system are presented in Table 4.14. A higher rank total indicated a lesser degree of perception of importance of linkage activities.

To determine whether the rank totals of the different sub-systems differ significantly among themselves, the Friedman non-parametric two-way analysis of variance test was applied. The results of the analysis revealed that all the $\chi^2 r$ values were significant indicating considerable difference among the sub-systems about the importance of performance of linkage activities as perceived by them.

Multiple comparisons among the different sub-systems were made on the basis of the sum of the ranks for perception of importance of each sub-system and the relevant critical ranges were computed using the procedure of Newman (1939) and Keuls (1952) extended to ordinal data (Zar, 1981).

4.4.1 *RSS with ESS, CSS and ISS*

From Table 4.14 it could be seen that the RSS perceived highly important linkage activities with CSS (88) followed by ESS (92) and ISS 132 in the descending order. It is also evident from the table that the perception of importance of linkage of RSS with ESS was not on par with that of CSS and ISS. Therefore, the general hypothesis set for the study that the perception about the importance of linkage activities of the RSS with ESS would be on par with that of CSS and ISS (3.9.2.1) is rejected. The related hypotheses that the perception about the importance of linkage activities of RSS with CSS would be on par with that of ESS and ISS (3.9.2.2) and that the perception about the importance of linkage activities of RSS with ISS would be on par with that of ESS and CSS (3.9.2.3) are also rejected.

TABLE 4.14: Perception about the importance of linkage activities between and among the different sub-systems-results of Friedman two-way non-parametric analysis of variance test

Sub-system	Sum of the ranks of perception				χ^2_r	SE	Critical range	
	RSS	ESS	CSS	ISS			W_2	W_3
Research N = 52	—	92	88	132	22.77*	7.21	20.19	24.23
Extension N = 134	358	—	217	229	91.21*	11.57	32.41	38.89
Client N = 110	256	145	—	259	76.74*	10.49	28.97	34.62
Input N = 32	76	49	67	—	11.81*	5.66	16.00	19.23

* Significant at 5 per cent level of probability

Conclusions :

RSS	a* CSS	a ESS	b ISS
ESS	a CSS	a ISS	b RSS
CSS	b ESS	a RSS	a ISS
ISS	b ESS	a CSS	a RSS

* Systems having the same letter on the top do not differ significantly.

The results of the study clearly indicate that the RSS perceived the linkage activities with (CSS and ESS as most important, though numerically a very small difference was there (88 and 92 respectively). But the linkage activities with ISS were perceived to be least important showing very urgent need for bettering the linkage between the two sub-systems. These results are in agreement with the results of extent of linkage activities performed between and among RSS-ESS, CSS and ISS. Though the sums of ranks for extent of linkage between RSS and ISS were just reversed, the rank value for RSS-ISS linkage and perception of importance of this linkage were the same (Table 4.1).

In the light of these findings, it could be conclusively stated that the

RSS-ESS and RSS-CSS linkage should be further strengthened and RSS-ISS linkage requires a thorough streamlining. The effectiveness of TOT of improved rice varieties released by the KAU depends on the coordinated efforts of all the four sub-systems. Even if any one of the sub-systems neither perform the linking activities nor perceive them as important, the effectiveness of the TOT process will be limited to that extent.

4.4.2 *ESS with RSS, CSS and ISS*

The perception of importance of linkage activities of ESS was the highest with CSS (217) and this was closely followed by linkage with the ISS (229). The performance of the ESS perceived its linkage activities with the RSS as least important, the rank value being 358 (Table 4.14). The table also revealed that the perception of the ESS about the importance of linkage activities with the RSS was not on par with that of the CSS and the ISS.

In view of these results, the general hypothesis set for the study that the perception of importance of linkage activities of ESS with RSS would be on par with that of CSS and ISS (3.9.2.4) is rejected. The connected hypotheses that the perception of importance of linkage activities of ESS with CSS would be on par with that of RSS and ISS (3.9.2.5) and that the perception of importance of linkage activities of ESS with ISS would be on par with that of RSS and CSS (3.9.2.6) are also rejected.

It may be noted that the extent of linkage activities of the ESS was highest with the ISS and this was followed by CSS in the descending order (Table 4.1). The same respondents perceived the importance of these linkage activities as most important with CSS followed by ISS.

This seems to be a plausible view since the extension personnel are in regular contact with the clientele under the T & V system for the transfer of fortnightly messages on major cultivated crops. It also seems quite reasonable to infer that the strong linkages between ESS and CSS will have positive bearing on the linkage between ESS and ISS because the recommendations of technology should be supported by timely supply of inputs.

4.4.3 *CSS with RSS, ESS and ISS*

The sum of the ranks for perception of importance of linkage of the

CSS with the ESS was 145 indicating the perception of linkage as most important (Table 4.14). The sums of ranks for linkage with the RSS and the ISS were 256 and 259 respectively indicating the performance of linkage as lesser important as perceived by the CSS. It could also be seen that the perception of importance of linkage activities of the CSS with the RSS was not on par with that of the ESS and ISS, whereas, the importance of linkage with the RSS was almost on par with that of ISS.

In the light of the above findings, the general hypothesis set for the study that the perception of importance of linkage activities of CSS with RSS would be on par with that of ESS and ISS (3.9.2.7) is rejected. Consequently, the related hypotheses that the perception of importance of linkage activities of CSS with ESS would be on par with that of RSS and ISS (3.9.2.8) and that the perception of importance of linkage activities of CSS with ISS would be on par with that of RSS and ESS (3.9.2.9) are also rejected.

Now it could be stated that, while the RSS perceived the linkage activities with CSS as most important, the CSS attaches highest importance for linkage with the ESS. The RSS and ISS were looked upon by the CSS almost on the same plane indicating the need for bettering the linkage activities among the three sub-systems. Nevertheless, it was encouraging to find that the CSS and ESS mutually agreeing on the importance of linkage activities between them which again conclusively proved that the relations between these two sub-systems had substantially improved after the introduction of T&V system.

4.4.4 *ISS with RSS, ESS and CSS*

The sums of the ranks for perception of importance of linkage activities of the ISS indicated maximum importance for linkage with the ESS (49) and this was followed by the CSS and the RSS in the descending order of importance (67 and 76 respectively). Thus the ISS does not perceive that its linkage activities with the RSS should be on par with that of the ESS and the CSS. Similarly, there was no parity of perception for importance of linkages between the RSS and CSS.

In view of these results, the general hypothesis set for the study that the perception of importance of linkage activities of ISS with RSS would be on par with that of ESS and CSS (3.9.2.10) is rejected. Therefore, the related hypotheses that the perception of importance of linkage activities of ISS with ESS would be on par with that of RSS and CSS (3.9.2.11) and that the perception of importance of linkage activities of ISS with CSS would be on par with that of RSS and ESS (3.9.2.12) are also rejected.

The point that emerges out of the results of the study is the fact that the ISS though has better linkages with the ESS, the former stands far apart from the RSS and the CSS. Similar expressions have come out of the RSS and the CSS in respect of their linkages with the ISS indicating mutual agreement. If the ISS has to be considered as a side support in the framework of agricultural production (Matlock, 1985), it has to perform effectively the linkage roles expected of it. The ISS should be in constant touch with the RSS and collect information on new varieties and other technologies recommended by the KAU. These information in turn could be transferred to the CSS through a number of extension activities like demonstrations in farmers' fields, improving the farmer contact programmes through village adoption, making available soil testing and plant protection services to the farmers etc. More succinctly the ISS, instead of confining itself in boosting up sales, should evolve a strategy of extension programme through a judicious mix of customer service and business promotion.

The foregoing results reveal an intriguing tendency prevailing among the four sub-systems studied. The hypotheses that the perceptions of importance of linkage activities by every sub-system would be on par, have been summarily rejected in all the cases indicating differential perception of importance of linkage activities when a sub-system-wise analysis is made. This differential perception is hazardous in view of the necessity for treating each sub-system as equally important in the TOT process. This also calls for the conscientisation of the four sub-systems on the equal importance of every other sub-system for a holistic view of the TOT process.

4.5 Importance of linkage activities between and among research, extension, client and input sub-systems as perceived by the respondents – Activity-wise analysis

The mean perception score for importance of each linkage activity under the different sub-systems were worked out and ranked. The overall mean for all the linkage activities under each sub-system was also calculated. Taking the overall mean score as the cut-off point, the linkage activities perceived as more important and less important by the respondents were discriminated. Based on this, the linkage activities to which further emphasis has to be given were extracted and interpreted.

4.5.1 *The Research Sub-system*

4.5.1.1 *RSS with ESS*:– As is evident from Table 4.15 there were 24 identified linkage activities to be performed by the RSS with ESS for effective TOT of improved rice varieties released by the KAU. Linkage activity X_{10} secured the highest mean score of 2.28 against the maximum possible score of 3.00. Activity X_{24} secured the lowest mean score of 1.55. The overall mean score of the 24 linkage activities was 1.93. Taking all those activities which secured a mean perception score of 1.93 and above, it could be observed that activities X_{10}, X_1, X_6, X_5, X_{21}, X_{17}, X_3, X_{22}, X_9 and X_{20}, were perceived to be more important, in the descending order, for TOT of improved rice varieties. All these 10 activities could be described as pivotal for linking the RSS and ESS. Attending regional workshops of NARP and T&V (X_{10}), involvement in training extension personnel (X_1), participation in joint field visits (X_6) which secured the first three ranks will go a long way in TOT of improved rice varieties.

All those 14 activities which secured mean score of 1.92 and below are also important in strengthening the RSS-ESS linkage especially X_{13}, X_{14}, X_7, X_8, X_{19}, etc. Activities X_{23} and X_{24} viz. participation in the KAU rice variety release committee and State rice variety release committee respectively, though will link the RSS with the ESS, are performed only by those nominated by the Vice-Chancellor. Considering the exclusive nature and limited spectrum of participation of these two activities it could be convincingly concluded that the RSS should bestow more attention in performing activities X_{13}, X_{14}, X_7; X_8 and X_{19}, X_{15}, X_{11} X_{12} and X_{16}. In other words, of the 24 linkage activities, majority of them are not being perceived as very important by the RSS and therefore need special emphasis for effective RSS-ESS linkage.

4.5.1.2 *RSS with CSS*:– Here also 24 linkage activities were included (Table 4.16). The activity X_9 secured the maximum mean score of 2.36 against maximum possible score of 3.00. The overall mean for all the 24 activities was 1.90. Thus taking 1.90 as the cut-off point it could be seen that linkage activities X_9, X_{14}, X_{10}, X_2 and X_{22}: X_3, X_{23}, X_{11}, X_{12}, X_{13} and X_{24}, X_1, X_8 and X_5 were perceived to be more important in the descending order to be performed by the RSS with the

TABLE 4.15: Perception of importance of linkage activities by RSS with ESS

Sl. No.	Activity	Mean perception score	Rank
X_1	Involvement in training extension personnel	2.26	2
X_2	Serving as course director of training	1.69	22
X_3	Serving as resource person of training	2.05	7
X_4	Serving as consultant for organising training	1.84	18
X_5	Serving as resource person for diagnosing field problems	2.17	4
X_6	Participation in joint field visits	2.25	3
X_7	Serving as resource person on field days	1.92	12.5
X_8	Serving as resource person on farmers' days	1.90	15.5
X_9	Serving as resource person in monthly workshops	1.98	9
X_{10}	Attending regional workshops of NARP and T&V system	2.28	1
X_{11}	Involvement as implementing officer of lab-to-land programme	1.78	19.5
X_{12}	Involvement as implementing officer of village adoption programme	1.69	22
X_{13}	Participation in conducting farm trials	1.92	12.5
X_{14}	Visits to result demonstrations	1.92	12.5
X_{15}	Visits to minikit trials	1.86	17
X_{16}	Participation in exhibitions	1.69	22
X_{17}	Participation in extension seminars	2.06	6
X_{18}	Participation in extension workshops	1.92	12.5
X_{19}	Writing in popular publications	1.90	15.5
X_{20}	Writing in newspaper columns	1.94	10
X_{21}	Attending package of practices workshops	2.07	5
X_{22}	Attending mini package workshops	2.03	8
X_{23}	Participation in KAU rice variety release committee	1.78	19.5
X_{24}	Participation in state level rice variety release committee	1.55	24
	Over all mean:	1.93	

CSS in the TOT of improved rice varieties. All these activities would no doubt will build better functional linkages between the RSS and the CSS. The importance attached to performance of clarification of doubts of farmers on rice in person (X_9), taking visiting farmers to experimental plots (X_{14}) and clarification of doubts of farmers on rice through letters (X_{10}) particularly indicated the personal attention bestowed to by the researchers in the TOT of improved rice varieties. Activity X_{16}-serving in the advisory committee of national demonstrations emerged to be least important as perceived by the RSS. This again might be due to the fact that only senior scientists alone get nominated to the advisory bodies and naturally lower level scientists like Junior Assistant Professors (JAPs), Assistant Professors (APs) would not attach much importance to such an activity. But activities X_{20}, X_{17}, X_{18}, X_{15}, X_6, X_7, X_{21}, X_{19}, and X_4 which secured mean perception score of 1.53 and above but less than 1.9 (overall mean score) are also important for building better functional relations with the two sub-systems. The results indicate that the RSS has less interest in mass contact activities like X_6, X_7, X_{18}, X_{19}, X_{20}, X_{21} etc. for which the researchers should be enthused further.

TABLE 4.16: Perception of importance of linkage activities by RSS with CSS

Sl. No.	Activity	Mean perception score	Rank
X_1	Organising training for farmers	1.94	12.5
X_2	Involvement in training, as resource person	2.09	4.5
X_3	Visits to farm trials	2.07	6
X_4	Contribution of articles in newspaper columns	1.82	15
X_5	Contribution of articles in magazines	1.90	14
X_6	Contribution of articles in souvenirs	1.69	19
X_7	Contribution of articles in books for farmers	1.73	18
X_8	Participation in radio programmes for farmers	1.94	12.5
X_9	Clarification of doubts of farmers on rice in person	2.36	1
X_{10}	Clarification of farmers' doubts through letters	2.11	3
X_{11}	Involvement in organising farmers' days	2.03	8
X_{12}	Involvement in organising field days	2.01	10
X_{13}	Visits to farmers fields on request	2.01	10
X_{14}	Taking the visiting farmers to experimental plots	2.28	2
X_{15}	Serving in the regional interdisciplinary committee	1.67	20
X_{16}	Serving in the advisory committee of NDS	1.50	24
X_{17}	Involvement in frontline demonstrations	1.59	22
X_{18}	Liaison with local co-operative societies	1.60	21
X_{19}	Liaison with farmers' associations of the area	1.78	16
X_{20}	Contribution of lessons for correspondence course on rice	1.53	23
X_{21}	Participation in campaigns	1.76	17
X_{22}	Organising exhibitions in research stations	2.09	4.5
X_{23}	Organising group discussions of farmers	2.05	7
X_{24}	Participation in group discussions organised by farmers	2.01	10
	Over all mean	1.90	

4.5.1.3 *RSS with ISS*:– Table 4.17 reveals that out of the 23 linkage activities of the RSS with the ISS, X_{18}-involving input agencies in exhibitions organised at research stations secured the highest mean perception score of 1.96 against a maximum possible score of 3.00. This indicates that almost all the rice research stations under the KAU involved the ISS in the exhibitions organised by researchers in connection with field days or farmers' days. The overall mean score for all the 23 linkage activities was only 1.74. Taking this as the cut-off point it could be seen that the activities X_{18}, X_9, X_7, X_8, X_{19}, X_6 and X_{11}; X_2 and X_{12}; X_{13} and X_{14} were perceived as more important, in the descending order, for TOT of improved rice varieties. The remaining 12 activities viz. X_{10}, X_4, X_5, X_{16} and X_{20}; X_{17}, X_1 and X_{22}; X_{21} and X_{23}; X_{15} and X_3 were perceived as less important for linkage. A cursory look at the table also reveals that invariably all the activities secured a mean perception score of less than 2.00 indicating that in general the involvement of the ISS in the TOT process was perceived to be less important by the RSS. The activities which secured a mean perception score less than 1.74 (the overall mean score) included keeping contact with fertilizer dealers, pesticide dealers, credit institutions like nationalised banks, local cooperative societies etc. These are also essential for strengthening the RSS-ISS linkage. The linkage components of change agent system of any country include technology developers, change agents, technology users and technology adoption facilitators. ISS being a strong facilitating sub-system, TOT process can be accelerated only by encouraging the participation of input agencies in the chain of linkage. The results, therefore, imply the need for encouraging the participation of the ISS in the TOT process and its contribution to overall efforts. This also calls for developing a strategy for active participation of the ISS in TOT at the policy making level.

4.5.2 *The Extension Sub-system*

4.5.2.1 *ESS with RSS:*– It could be seen from Table 4.18 that out of the 21 linkage activities of the ESS with the RSS, activity X_8 ranked first with mean perception score of 2.42 out of the maximum possible score of 3.00 which indicated that there was good participation of extension personnel in the field days organised at research stations. The overall perception score for all the linkage activities was 2.11. It could also be found that 11 out of 21 activities secured mean perception scores well above the overall mean score. The other activities perceived as most important were giving suggestions for taking up new area of research on rice (X_{20}), joint field visits with researchers (X_7), visits to the experimental plots of rice at research stations (X_1), cooperating with researchers in laying out farm trials

(X_2) etc. in the descending order. These will definitely strengthen the ESS-RSS linkage. This could reasonably be attributed to the impact of the interactions between the two sub-systems in the monthly workshops of T&V system and combined regional workshops of T&V and NARP.

But the fact remains that some other linkage activities like X_3, X_4, X_5, X_{12}, X_{13}, X_{14}, X_{15} and X_{16} which secured less than the overall mean score of 2.11 were perceived to be less important by the ESS, eventhough they are equally important in building effective linkage. Only the senior scientists participate in pre-monthly and monthly workshops of T&V system, the zonal workshops of T&V and NARP and state level package of practices workshops. Since a good majority of other scientists are deprived of the opportunity of participating in these important workshops their discontent might have been reflected in the poor perception score. Drawing a cross section of scientists to participate in such workshops would perhaps be a corrective step.

TABLE 4.17: Perception of importance of linkage activity by RSS with ISS

Sl. No.	Activity	Mean perception score	Rank
X_1	Keeping contact with fertilizer dealers	1.65	18.5
X_2	Keeping contact with pesticide dealers	1.78	8.5
X_3	Keeping contact with local nationalised banks	1.51	23
X_4	Keeping contact with local co-operation societies	1.71	13
X_5	Conducting trials with fertilisers supplied by dealers	1.69	15
X_6	Conducting trials with pesticides supplied by dealers	1.80	6.5
X_7	Inviting fertiliser dealers in seminars at research stations	1.92	3
X_8	Inviting weedicide dealers for seminars at research stations	1.88	4
X_9	Inviting pesticide dealers for seminars at research stations	1.94	2
X_{10}	Inviting credit agencies for seminars at research stations	1.73	12
X_{11}	Participation in seminars organised by input agencies	1.80	6.5
X_{12}	Participation in workshop organised by input agencies	1.78	8.5
X_{13}	Feeding back the results of fertilizer trials to dealers	1.76	10.5
X_{14}	Feeding back the results of pesticide trials to dealers	1.76	10.5
X_{15}	Helping fertiliser dealers in laying out demonstrations	1.57	22
X_{16}	Involving input agencies in lab-to-land programme	1.69	15
X_{17}	Involving input agencies in village adoption programme	1.67	17
X_{18}	Involving input agencies in exhibitions	1.96	1
X_{19}	Involving input agencies in farm trials	1.84	5
X_{20}	Involving input agencies in NDS	1.69	15
X_{21}	Contributing articles in publications of input agencies	1.61	20.5
X_{22}	Helping input agencies as consultants	1.65	18.5
X_{23}	Keeping contact with input agencies through correspondence	1.61	20.5
	Over all mean	1.74	

TABLE 4.18: Perception of importance of linkage activities by ESS with RSS

Sl. No.	Activity	Mean perception score	Rank
X_1	Visiting experimental plots at rice research stations	2.36	4
X_2	Cooperating with researchers in laying out farm trials	2.35	5
X_3	Participation in pre-monthly workshops of T & V system	2.04	14
X_4	participation in monthly workshops of T & V system	1.91	17
X_5	Participation in zonal workshops of NARP and T & V system	1.76	21
X_6	Taking field problems on rice to researchers	2.23	7
X_7	Participation in joint field visits with researchers	2.39	3
X_8	Participation in field days organised by researchers	2.42	1
X_9	Attending package of practices workshops	1.84	19
X_{10}	Involving researchers in seminars	2.14	10
X_{11}	Taking farmers to rice research stations	2.18	8
X_{12}	Taking help of researchers in laying out minikit trials	2.05	13
X_{13}	Taking help of researchers in laying out demonstrations	1.95	16
X_{14}	Involving researchers in exhibitions	1.82	20
X_{15}	Participation in exhibitions at research stations	1.85	18
X_{16}	Participation in workshops organised by researchers	1.97	15
X_{17}	Participation in rice production technology training at research stations	2.16	9
X_{18}	Participation in special training on rice at research stations	2.26	6
X_{19}	Involvement in campaigns organised by researchers	2.11	11
X_{20}	Giving suggestions to take up new areas of research on rice	2.41	2
X_{21}	Cooperating with researchers in lab-to-land programme	2.10	12
	Over all mean	2.11	

4.5.2.2 *ESS with CSS*:– A perusal of Table 4.19 reveal that the linkage activity X_{12}-arranging for timely supply of material inputs to farmers emerged with highest perception score of 2.64 out of the maximum possible score of 3.00. It is also evident from the table that 13 activities out of the total 19, secured overall mean perception score of 2.27 and above. They were X_{12}, X_{11}, X_1, X_2, X_4, X_{13}, X_{18}, X_{17}, X_{19}, X_{16} and X_3; X_{16} and X_{10} in the descending order. The results are gratifying because the ESS seems to be fully conscious of their responsibility of arranging for timely supply of inputs (X_{12}), visiting the farmers' fields on request (X_{11}), visiting regularly paddy fields of contact farmers (X_1), visits to farm trials on rice (X_2) etc. Adoption of any new technology depends upon making available the material component if any, supplemented by constant visits to give recommendations at various stages of adoption. In this study these aspects have been fully looked into the ESS and hence a better perception score for these activities. But what matters is the low mean perception score (<2.27) secured by some linking activities through mass media. These include X_{14}, X_8, X_5, X_9, X_6, and X_7. The results suggest that the ESS should involve themselves in organising exhibitions. writing in newspaper columns and popular publications, providing information through All India Radio etc. to a great extent. Perhaps lack of expertise in contributing to mass media programmes might be holding them back from such participation.

TABLE 4.19: Perception of importance of linkage activity by ESS with CSS

Sl. No.	Activity	Mean perception score	Rank
X_1	Visiting to paddy fields of contact and other farmers	2.56	3
X_2	Visits to farm trials on rice	2.55	4
X_3	Involvement in organising farmers' day	2.30	10.5
X_4	Organising farmers' group discussions on rice	2.53	5
X_5	Inviting farmers for seminars	1.98	16
X_6	Contribution of articles on rice in newspaper columns	1.70	18
X_7	Contribution of articles on rice in popular publications	1.68	19
X_8	Providing information support to farmers through All India Radio	2.05	15
X_9	Contributing articles to publications of farmers' association	1.89	17
X_{10}	Conducting demonstrations on rice	2.27	13
X_{11}	Visits to farmers' fields on request	2.58	2
X_{12}	Arranging for timely supply of material inputs	2.64	1
X_{13}	Making available cash input through credit agencies	2.51	6
X_{14}	Organising exhibition on rice for farmers	2.20	14
X_{15}	Organising Campaigns on rice	2.28	12
X_{16}	Organising training on rice production technology for farmers	2.30	10.5
X_{17}	Keeping contact with local farmers associations	2.35	8
X_{18}	Attending the discussions organised by farmers	2.48	7
X_{19}	Giving recognition to innovative farmers	2.33	9
	Over all mean	2.27	

4.5.2.3 *ESS with ISS:–* A glance at Table 4.20 indicates that out of the 16 linkage activities, 10 activities secured the overall mean perception score of 2.00 and above. They were X_1, X_2, X_4 and X_8, X_9, X_3, X_7, X_{12}, X_5 and X_{16} in the descending order. Activity X_1 had the highest individual mean perception score followed by X_2, indicating the best efforts put in by the ESS in liaising with the ISS for making available material and cash inputs in time. Activities which secured mean perception score of <2.00 were X_{10}, X_{11}, X_6, X_7 and X_{13}; X_{15} and X_{14}. Of these, X_{14} and X_{15} could be performed only if the demand comes from the ISS itself. Participation in seminars and exhibitions organised by the ISS (X_{11}) and (X_{13}) again calls for the initiative from the ISS and the frequency of organising such seminars and exhibitions is very less. Similarly, though the input agencies are also expected to be invited on the SMS training days (X_6) it is never done and hence perceived to be less important. X_7-checking the quality of fertilizers/pesticides is a duty of the AOs only and naturally other categories of respondents of the ESS could perceive it as less important. However, contribution of articles to publications of the ISS and inviting the input agencies on SMS training days could be regular feature and also will help to strengthen further the ESS-ISS linkage. On the whole, the perception of the ESS about the importance of linkage activities with the ISS was found to be satisfactory and encouraging.

TABLE 4.20: Perception of importance of linkage activities by ESS with ISS

Sl. No.	Activity	Mean perception score	Rank
X_1	Arranging for supply of material inputs during seasons	2.50	1
X_2	Making available cash inputs to farmers in time	2.35	2
X_3	Inviting input agencies for farmers' days	2.27	6
X_4	Involving input agencies in seminars	2.29	3.5
X_5	Involving input agencies in exhibitions	2.06	9
X_6	Inviting input agencies on SMS training days	1.89	12.5
X_7	Checking the quality of fertilizers/pesticides	2.22	7
X_8	Taking steps to open sales points wherever required	2.29	3.5
X_9	Keeping touch with input agencies for latest information	2.28	5
X_{10}	Contributing articles to publications of input agencies	1.74	16
X_{11}	Participation in seminars organised by input agencies	1.94	11
X_{12}	Participation in campaigns organised by input agencies	2.07	8
X_{13}	Participation in exhibitions organised by input agencies	1.89	12.5
X_{14}	Organising training for input agencies	1.75	15
X_{15}	Providing consultancy service to input agencies	1.78	14
X_{16}	Conveying complaints of farmers to input agencies	2.00	10
	Over all mean	2.00	

4.5.3 *The Client Sub-system*

4.5.3.1 CSS with RSS:– Table 4.21 presents the mean perception scores of linkage activities performed by the CSS with the RSS. Activity X_3-contacting researchers for specific recommendations on rice secured the highest mean score of 2.42 which is infact one of the most important linking activities of the CSS. This activity gained much significance particularly after the popularisation of improved rice varieties. The overall mean perception score for all the 17 linkage activities was 2.26, and 10 out of 17 activities crossed this value. They included X_3, X_4, X_{15}, X_{12}, X_{10}, X_{13}, X_5 and X_7; X_{16} and X_8 in the descending order. All these activities are of vital importance for building up of strong linkage between the CSS and the RSS and hence the results are encouraging. Among the activities which had less than the overall mean score of 2.26 were X_{14}, X_{11}, X_1 and X_2; X_6, X_{17} and X_9 in the descending order. It could be found that making correspondence with the research station (X_9) was the item perceived to be the least important. This can be attributed to the cost involved in frequent correspondence. Similarly activities like visits to research stations (X_1), participation in training organised at research stations (X_2) participation in farmers' day (X_6), visits to trial plots on other farmers' fields (X_{14}) and reading articles in popular publications were perceived to be lesser important though these activities will add strength to the CSS-RSS linkages. Distance from the place of residence, prior engagements etc. may be the reasons for poor perception of activities X_1, X_2 and X_{14}. There are only two popular publications viz. Kalpadhenu of the KAU and Keralakarshakan of the Farm Information Bureau, Government of Kerala. These are priced periodicals and many of the respondent farmers might not have enrolled themselves as subscribers and hence perceived as less important. Similarly the benefits of village adoption programme are extended to only a selected few and the reason for perceiving less important could be attributed to this. However, every effort should be made to make them visit the research stations and participate in trainings organised by research stations which will be immensely beneficial for an on-the-spot study of newer varieties and interacting with scientists and thereby strengthening the relations between the two sub-systems.

TABLE 4.21: Perception of importance of linkage activities by CSS with RSS

Sl. No.	Activities	Mean perception score	Rank
X_1	Visits to rice research station	2.21	13.5
X_2	Participation in training organised by rice research stations	2.21	13.5
X_3	Contacting researchers for specific recommendations on rice	2.42	1
X_4	Participation in Kisan melas organised by rice research stations	2.40	2
X_5	Participation in exhibitions organised by rice research stations	2.28	7.5
X_6	Participation in farmers' day organised by rice research stations	2.20	15
X_7	Obtaining seeds from rice research stations	2.28	7.5
X_8	Cooperating with researchers in the conduct of trials	2.26	10
X_9	Making correspondence with rice researchers	1.89	17
X_{10}	Reading the articles on rice in newspapers	2.35	5
X_{11}	Reading the articles on rice in popular publications	2.24	12
X_{12}	Listening to the talk of researchers through AIR	2.37	4
X_{13}	Giving land for conducting trials by researchers	2.30	6
X_{14}	Visits to the trials in other farmers' fields	2.25	11
X_{15}	Seeking help of researchers to solve problems on rice	2.38	3
X_{16}	Getting benefits (inputs) through lab-to-land programmes	2.27	9
X_{17}	Getting benefit (inputs) through village adoption programme	2.13	16
	Over all mean	2.26	

4.5.3.2 *CSS with ESS*:– Data in Table 4.22, reveal that two linkage activities X_5-participation in seminars organised by extension workers and X_7-seeking help of extension workers in solving problems on rice had emerged as equally important. These activities secured the highest mean perception score of 2.53 against the maximum possible score of 3.00. These two activities are important from the point of view of linkage through direct interactions with the extension personnel. The overall mean score for the 16 activities was 2.35. In addition to X_5 and X_7, activities X_2, X_8, X_4, X_{13} and X_{14}; X_{11}, X_9 and X_3 were found to have crossed this limit indicating that the CSS had perceived these activities important for building up linkages with the ESS. It could also be seen that there were six activities viz. X_6, X_{12}, X_{16}, X_{10}, X_1 and X_{15} which secured a score of less than 2.35, in the descending order. These included listening to the talk given by extension personnel through AIR, inviting extension personnel in discussions, reading articles on rice in popular publications, participation in trainings organised by extension personnel and correspondence with extension personnel. Though these are important from the point of view of establishing linkage between the CSS and the ESS, the CSS did not give much importance to them. Therefore, the results suggest that if these activities also were given due importance by the CSS, the existing linkages could be further cemented.

TABLE 4.22: Perception of importance of linkage activities by CSS with ESS

Sl. No.	Activities	Mean perception score	Rank
X_1	Participation in training organised by extension workers	2.08	15
X_2	Cooperating in farm trials conducted by extension personnel	2.52	3
X_3	Participation in demonstration laid out by extension personnel	2.37	10
X_4	Participation in farmers' day organised by extension personnel	2.45	5
X_5	Participation in seminars organised by extension personnel	2.53	1.5
X_6	Participation in exhibitions organised by extension personnel	2.32	11
X_7	Seeking help of extension workers for solving problems on rice	2.53	1.5
X_8	Participation in discussions organised by extension personnel	2.48	4
X_9	Reading articles in newspapers written by extension personnel	2.40	9
X_{10}	Reading articles on rice in popular publications	2.19	14
X_{11}	Visiting the agricultural extension units	2.43	8
X_{12}	Listening to the talk given by extension personnel through All India Radio	2.29	12
X_{13}	Contacting the extension personnel for inputs	2.44	6.5
X_{14}	Sharing problems/experiences with extention workers	2.44	6.5
X_{15}	Correspondence with extension workers	1.88	16
X_{16}	Inviting extension personnel in discussions	2.24	13
	Over all mean	2.35	

4.5.3.3 *CSS with ISS*:– The mean perception scores of linkage activities of the CSS with the ISS are presented in Table 4.23. A perusal of the table reveals that out of the 12 linkage activities, feeding back to the ISS about quality of inputs (X_{11}) ranked first by securing the highest mean perception score of 2.31 against a maximum possible score of 3.00. This evidently showed the concern of the CSS about the availability of quality inputs for their rice crop, and perhaps the existing linkage could be further improved by ensuring supply of quality inputs commensurate with increased cost. When the overall mean score of 2.10 was taken as the cut-off point, activities X_{11}, X_7,

X_{12}, X_6, X_8 and X_{10} crossed the limit and thus the performance of these activities was perceived more important by the CSS. Simultaneously, another six activities viz. X_2, X_1 and X_3, X_4, X_5 and X_9 were perceived to be less important in the descending order. It could be seen that these included some important linking functions like participation in seminars organised by the input agencies, collecting technical information from them, participation in discussions organised by the ISS; visiting their exhibitions and cooperating with the ISS in conducting exhibitions. The results point to the comparatively poor linkage existing between these two sub-systems. It can be concluded that measures for building up a much strong functional relationship with the CSS should be initiated by ISS.

TABLE 4.23: Perception of importance of linkage activities by CSS with ISS

Sl. No.	Activities	Mean perception score	Rank
X_1	Collecting technical information from input agencies	2.02	8.5
X_2	Participation in seminars organised by input agencies	2.05	7
X_3	Participation in discussions organised by input agencies	2.02	8.5
X_4	Visiting exhibitions organised by input agencies	1.98	10
X_5	Cooperating with input agencies in demonstrations	1.94	11
X_6	Visiting the result demonstrations of input agencies	2.14	4
X_7	Availing credit facilities from credit agencies	2.19	2
X_8	Trying to know about the services available with input agencies	2.13	5
X_9	Taking help of input agencies in preparing production plan	1.80	12
X_{10}	Reading the publications of input agencies	2.11	6
X_{11}	Feeding back about quality of inputs	2.31	1
X_{12}	Feeding back to input agencies about their performance	2.18	3
	Over all mean	2.10	

4.5.4 *The input sub-system*

4.5.4.1 *ISS with RSS*:– Data in Table 4.24 show that participation in field days organised by researchers (X_4) emerged as the most important linkage activity as perceived by the ISS, securing the highest mean perception score of 2.43 out of the possible maximum score of 3.00. Taking the overall mean score 2.0 as cut-off point, it could be seen that out of the total 13 activities, only six including X_4 secured the score of

2.0 and above. True that these included some very important linking activities such as participation in field days, exhibitions, package of practices workshops etc. organised by researchers. But more than 50 per cent of the linkage activities viz. X_2 and X_9; X_7, X_{11} and X_{13}; X_{10} and X_{12} were found to be perceived as less important linkage activities for TOT of improved rice varieties. Getting new fertilizers/pesticides tested through yield trials at rice research stations (X_2 and X_{11}) perhaps could be construed as very constructive linkage activities. But these were perceived to be less important. The possible reason would be the enhanced rate of fees to be paid by the ISS to the KAU for undertaking such field trials. Some other activities like mailing publications to researchers (X_7), taking advice of researchers in laying out demonstrations of input agencies (X_{10}) etc. could be given more importance by the ISS so that the functional relations between the ISS and the RSS could be further strengthened. Data on the whole, suggest the need for further strengthening of the ISS-RSS linkage for effective TOT of improved rice varieties.

TABLE 4.24: Perception of importance of linkage activities by ISS with RSS

Sl. No.	Activities	Mean perception score	Rank
X_1	Keeping contact with scientists of rice research stations	2.21	3.5
X_2	Getting new fertilisers/pesticides tested at research stations	1.96	7.5
X_3	Collecting information on new rice varieties released by KAU	2.12	5
X_4	Participation in field days organised by researchers	2.43	1
X_5	Participation in exhibitions at rice research stations	2.31	2
X_6	Participation in package of practices workshop	2.00	6
X_7	Mailing publications to rice research stations	1.90	9
X_8	Inviting researchers in melas organised by input agencies	2.21	3.5
X_9	Feeding back to researchers about field experiences	1.96	7.5
X_{10}	Taking advice of researchers in laying out demonstrations	1.81	12
X_{11}	Giving fertilisers/pesticides to research stations for trials	1.87	10.5
X_{12}	Extending financial help to research stations for melas/field days	1.78	13
X_{13}	Contributing articles to research publications	1.87	10.5
	Over all mean	2.00	

4.5.4.2 *ISS with ESS*:– Table 4.25 depicts the mean perception scores of importance of linkage activities of the ISS with the ESS. The highest score of 2.59 against a maximum possible score of 3.00 was secured by activity X_1 viz. participation in campaigns organised by extension agencies. The overall mean score for all the 18 activities was 2.15. It could also be seen that out of the 18 activities, nine had mean scores 2.15 and above. The results indicated that the ISS was keen to participate in campaigns, melas and seminars organised by the extension agencies and keeping in touch with them (X_1, X_2, X_3, X_5 and X_{10} respectively). Similarly, the ISS also involves the ESS in campaigns and seminars organised by them (X_8 and X_7). Other activities perceived as important were participation in exhibition and dealers training organised by extension agencies (X_4 and X_{11}). All these activities, no doubt are crucial for effective linkage and they are being perceived as very important too.

Linkage activities perceived as less important (overall mean score <2.15) were X_{14}, X_9, X_{13} and X_{15}, X_{16}, X_6 and X_{17}; X_{18} and X_{12} in the descending order. Of these, X_{12} was performed only by a few input agencies and hence ranked the last. If the ISS could bestow some more attention to the performance of the remaining eight activities also, the linkage would have been stronger than what it is at present.

TABLE 4.25: Perception of importance of linkage activities by ISS with ESS

Sl. No.	Activities	Mean perception score	Rank
X_1	Participation in campaigns organised by extension agencies	2.59	1
X_2	Participation in melas organised by extension agencies	2.56	2
X_3	Participation in seminars organised by extension agencies	2.53	4
X_4	Participation in exhibitions organised by extension agencies	2.25	8
X_5	Keeping the extension agencies informed about latest products	2.53	4
X_6	Mailing literature to extension personnel	1.84	15.5
X_7	Inviting extension personnel for seminars	2.31	7
X_8	Inviting extension personnel for campaigns	2.34	6
X_9	Taking help of extension agencies for tackling field problems	2.00	11
X_{10}	Keeping constant contacts with extension personnel	2.53	4
X_{11}	Attending dealers training organised by SDAOs	2.21	9
X_{12}	Organising special training for students of agriculture	1.56	18
X_{13}	Attending fortnightly training organised by SDAOs	1.96	12.5
X_{14}	Cooperating with KAU in lab-to-land programme	2.06	10
X_{15}	Cooperating with KAU in village adoption programme	1.96	12.5
X_{16}	Cooperating with extension personnel in farm trials	1.93	14
X_{17}	Financing extension agencies for seminars/workshops	1.84	15.5
X_{18}	Cooperating with KAU in national demonstrations	1.78	17
	Over all mean	2.15	

4.5.4.3 *ISS with CSS*:– The mean perception scores for linkage activities of the ISS with the CSS are presented in Table 4.26. Timely supply of inputs to farmers (X_1) was ranked 1st with the highest mean score of 2.62 against the possible maximum score of 3.00. The overall mean score for all the 19 linkage activities was 2.15. Taking all the linkage activities with overall mean score of 2.15 and above, it could be seen that majority (13 out of 19) of the activities were perceived by the ISS as more important functions for linking with the CSS. These activities included X_1, X_8, X_{12}, X_{15}, X_2, X_6 and X_9; X_5, X_7, X_{10} and X_{14}; X_4 and X_3 in the descending order. All these activities are crucial for building up a strong linkage between the two sub-systems and it was encouraging that the ISS perceived them as most important activities, eventhough the extent to which these were performed might vary.

The remaining six activities viz. X_{19}, X_{16}, X_{18} and X_{17}; X_{13} and X_{11} which were perceived as less important are also, infact, important for strengthening the linkages, especially the activity X_{19}-special farmer contact programme. Similarly activities X_{16}, X_{17} and X_{18} could not be performed by all because only fertilizer dealers undertake soil testing services, and dealers of pp chemicals and equipment could make available pp equipment to farmers. Activity X_{18}-helping farmers in marketing rice does not come under the purview of the above two agencies, but mostly a function of marketing cooperative societies. Therefore the reason for perceiving these activities as less important could be attributed to this selective and specialised nature of activities of the ISS. In general, the perception has been quite satisfactory.

4.6 Constraints in evolving, spreading and adopting improved rice varieties released by the KAU as perceived by RSS, ESS, CSS and ISS

As stated elsewhere, the spread of improved rice varicties in Kerala is around 25 per cent of the total area under rice even after one and a half decade of the introduction of the high yielding varieties. This must be due to certain limitations with individual varieties in respect of their suitability to agro-climatic conditions, tolerance to pests and diseases, grain quality, straw yield, cost-benefits etc. All these limitations put together can be termed as constraints. It will be of great use to identify these constraints as perceived by the different sub-systems. It will also help to know how far the different sub-systems agree with each other in perceiving the importance of these constraints. Hence, an attempt was made to identify the constraints in evolving and spreading improved rice varieties as perceived by the RSS and the constraints in spreading and adopting these varieties as perceived by the ESS, CSS and ISS. The results in this regard are present hereunder.

4.6.1 *Constraints in evolving and spreading improved rice varieties as perceived by the RSS*

The role of the RSS does not end by evolving and releasing for cultivation the improved rice varieties. Researchers should be open to receive the feedback about the performance of these varieties under field conditions.

Table 4.27 presents the constraints in evolving and spreading improved varieties as perceived and ranked by the respondents of the RSS.

TABLE 4.26: Perception of importance of linkage activities by ISS with CSS

Sl. No.	Activities	Mean perception score	Rank
X_1	Timely supply of inputs to farmers	2.62	1
X_2	Laying out demonstrations in farmers' fields	2.21	6
X_3	Opening sales point at required places	2.15	13
X_4	Adoption of villages for farmer contact programme	2.16	12
X_5	Organising training on rice for farmers	2.20	8
X_6	Organising campaigns	2.21	6
X_7	Participation in input fortnight celebrations	2.18	10
X_8	Organising group discussions of farmers	2.37	2
X_9	Organising exhibitions for farmers	2.21	6
X_{10}	Supplying printed information materials to farmers	2.18	10
X_{11}	Arranging film shows on rice production technology	1.78	19
X_{12}	Helping farmers in availing credit facilities	2.28	3
X_{13}	Publication of journals/newsletters for farmers	1.90	17.5
X_{14}	Providing technical information to farmers on request	2.18	10
X_{15}	Receiving complaints of farmers properly	2.25	4
X_{16}	Extending soil testing service to farmers	2.03	15
X_{17}	Making available PP equipment to farmers	1.90	17.5
X_{18}	Helping farmers in marketing rice	1.96	16
X_{19}	Organising special farmer contact programme	2.09	14
	Over all mean	2.15	

Taking the first four constraints into consideration the following observations could be made. Rice cultivation in Kerala has its own special features due to varied soil, land and physiographic and climatological factors. Therefore, the future of the improved rice varieties would seem to depend upon fresh efforts and breeding varieties to suit the local conditions. It is in this context, the first constraint has to be viewed. The fact that in spite of all efforts put in by researchers, extension workers and input agencies, the coverage of HYV in Kerala is only around 25 per cent which suggests the need for evolving a breeding programme for varieties suited to dissimilar rice growing tracts.

TABLE 4.27: Constraints in evolving and spreading improved rice varieties released by the KAU as perceived by RSS

n = 52

Sl.No.	Constraints	Ranks
1.	Lack of facilities for evolving specific varieties suited to certain rice growing pockets of Kerala	I
2.	Non-filling up/transfer of scientists makes it difficult to take up new items of crop improvement	II
3.	Lack of facilities to multiply seeds of improved varieties to meet the demand	III
4.	Lack of facilities for on-farm teaching of scientific management practices and follow-up	IV
5.	Improved rice varieties performing in one area may not be performing similarly in other areas	V
6.	Rice is cultivated under a variety of soil and field conditions. Improved rice varieties suited to all such conditions are not available	VI
7.	Majority of farmers are not full time rice farmers and hence rice cultivation is not taken as a business	VII
8.	It is difficult to combine high yielding and other desired characteristics in one and the same variety	VIII
9.	Evolving new variety is a laborious and time consuming process	IX

The constraint ranked as second points to the administrative lacuna. Many posts of researchers are kept vacant over a prolonged period and research personnel are being transferred frequently which lead to inordinate delay in the execution of research programmes.

Regarding the lack of facilities to multiply improved rice varieties to meet the demand of farmers, it appears to be a genuine problem. This problem was posed especially at agricultural sub-division, Alwaye while researcher was interviewing the respondent farmers.

Vyttila-3, an improved rice variety released by RRS Vyttila has gained popularity among the farmers, but the major constraint was non-availability of sufficient quantity of the seeds and hence farmers were forced to stick on to the traditional varieties. Steps need be initiated to find a solution to this problem, lest the entire efforts on strengthening research-extension linkages would go a waste.

With regard to the fourth constraint, the link between researcher-client sub-systems will not be effective unless, the scientists get themselves exposed to the field problems of the rice growers. The research sub-system could link itself with the client sub-system by devoting part of their time in training selected farmers in the scientific techniques of growing improved rice varieties. Multi-locational trials could be increased to ensure enough feedback. Similarly, the rice researchers must involve themselves more in the TOT process like national demonstrations, lab-to-land programme and village adoption programme.

4.6.2 *Constraints in spreading and adopting improved rice varieties as perceived by extension, client and input sub-systems*

The constraints in spreading and adopting improved rice varieties released by the KAU, selected based on relevancy weightage were again got ranked by the concerned sub-system and the results are presented in the Table 4.28.

TABLE 4.28: Constraints in spreading and adopting improved rice varieties released by KAU as perceived by extension, client and input sub-systems

Sl.No.	Constraints	Rank
1.	Improved rice varieties are found to be more susceptible to pests and diseases	I
2.	Improved rice varieties require heavy dose of fertilisers and timely plant protection measures	II
3.	Poor cooking quality of improved rice varieties discourage farmers to adopt improved rice varieties	III
4.	Straw yield of improved rice varieties is comparatively low	IV
5.	Low market value is one of the constraints in the wider adoption of improved rice varieties	V
6.	Weed problem is more when improved rice varieties are grown	VI
7.	Improved rice varieties are not tolerant to unfavourable soil conditions	VII
8.	Under same management conditions yield of local varieties is better than improved rice varieties	VIII
9.	Want of assured irrigation and drainage facilities stand in the way of adopting improved rice varieties	IX

Taking the first four constraints in spreading and adopting improved rice varieties released by the KAU, as perceived by the ESS, CSS and ISS for discussion, it could be seen that the constraints I and II viz. susceptibility to pests and disease and requirement of heavy dose of fertilizers and timely plant protection measures very well conform to the results of the study conducted by Panikar (1981) in Kuttanad and Palghat, the two major rice growing areas of Kerala. It is true that the improved rice varieties could trigger off a vertical growth in productivity, but the problems of pests and diseases and rising cost of fertilizers and chemicals pushed up the cost of production. As a result of this, paddy cultivation has come down to the lower rung of the priority ladder. Since increased costs of production are not capable of being reduced by themselves, the possible remedial measures seem to be the re-introduction of pests and diseases surveillance, introduction of custom hire service of plant protection equipment and promotion of the use of balanced and slow release fertilizers.

The third constraint viz. poor cooking quality of improved rice varieties, arises out of the preference of Keralites to red and coarse grains and hence varieties which do not satisfy these requirements are not likely to be accepted by the farming community. As a matter of fact, majority of the varieties released by the KAU were red rice except 'suvarnamodan' and 'suvarnaprabha' which are suited to certain specific location like the uplands. Hence, this constraint does not appear to be much relevant to the KAU rice varieties. But the respondents perhaps would have in mind the other HYVs like Jaya and IR-8 two white bran varieties having low consumer preference.

The fourth constraint – low straw yield is equally important because in Palghat district, straw has a dual use — as a fodder and a thatching material for houses of low income group. An overview of the extent of cultivation of improved rice varieties released by the KAU would reveal that the variety Jyothi has maximum coverage (20.60 per cent of the total rice area) in all the eight sub-divisions under study. This could be attributed to the fact that Jyothi is a medium tall red rice variety and hence it satisfies the requirements of both straw yield and grain quality, in comparison to other popular red rice varieties like Bharathi, Sabari, Pavizham etc. which are dwarf in stature.

The results of the study bring into focus one significant point. Any improved rice variety should satisfy the consumer requirements. Endorsing the view of researchers that it is difficult to combine all the desirable characteristics in one and the same variety, the future of improved rice varieties would seem to depend on evolving a research programme of breeding new varieties to suit the local conditions.

4.7 Empirical model of the linkages between and among the research, extension, client and input sub-systems in the TOT of improved rice varieties released by the KAU

In the light of the results emerging from the Friedman two-way analysis of variance by ranks, the extent of linkage existing between and among the four sub-systems is presented in Fig. 4.16.

The system of TOT of improved rice varieties consists of four interrelated (double headed arrows) sub-systems, namely, the research, extension, client and input sub-systems. The study revealed that there was strong linkage (straight line arrows) between the ESS and the CSS, and between the ESS and the ISS and in both the cases, the linkages were reciprocal.

On the contrary, the RSS had no reciprocal strong linkage with any of the other sub-systems. Though the RSS perceived to have strong linkages with the ESS and the CSS; the ESS and the CSS found to have weak linkage (dotted line arrows) with the RSS.

Similarly, the ISS though it perceived to have strong linkage with the RSS and the CSS, these two sub-systems were found to have weak linkages with the RSS.

Thus the RSS was found to be the weakest link in functional linkage followed by the ISS in the TOT of improved rice varieties released by the KAU.

4.8 Suggested model of linkages between and among the research, extension, client and input sub-systems in the transfer of agricultural technology in Kerala

Flowing from the results of the study, a model of the linkages between and among the RSS, ESS, CSS and ISS in the transfer of agricultural technology in Kerala has been suggested wherein steps to strengthen the existing structural and functional linkages have illustrated in Fig. 4.17.

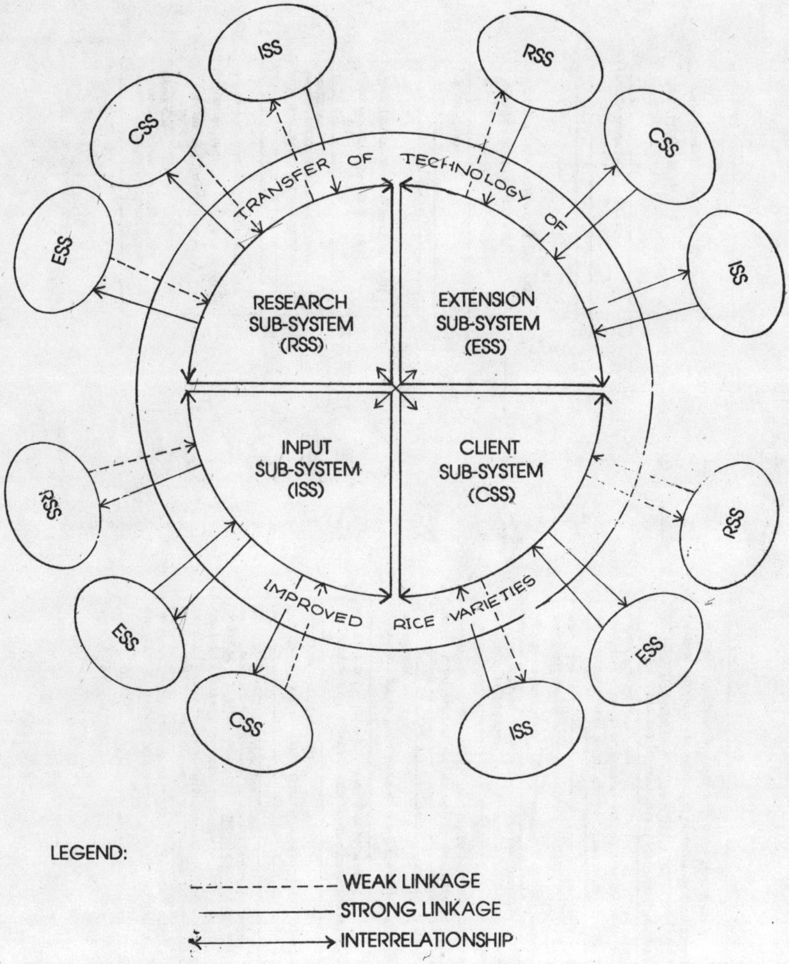

FIG. 4.16: Empirical model of the linkages between and among the research, extension, client and input sub-systems in the TOT of improved rice varieties released by KAU

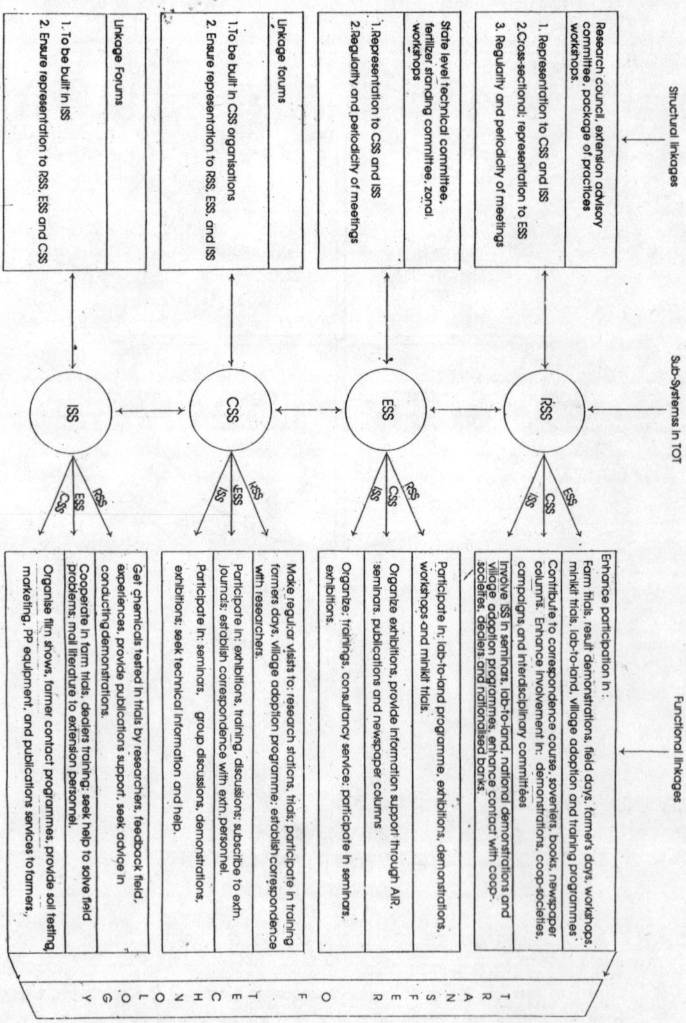

FIG. 4.17: Suggested model of linkages for transfer of agricultural technology

5

SUMMARY

EFFECTIVE TOT calls for well organised and coordinated interaction between and among the RSS, ESS, CSS and ISS. The spread of HYVs of rice in Kerala is only around 25 per cent of the total area under rice which could be due to the lack of effective linkages between and among the sub-systems involved in the TOT process. Therefore, with a view to study the existing structural and functional linkages between and among the four sub-systems in the TOT of improved rice varieties released by the KAU, the present study was taken up with the following specific objectives:

1) To study the extent of linkage activities performed by, between and among the research, extension, client and input sub-systems in the TOT of the improved rice varieties released by the KAU.
2) To analyse the factors associated with the linkages between and among the different sub-systems in the TOT of improved rice varieties released by the KAU.
3) To identify the constraints in evolving, spreading and adopting improved rice varieties as perceived by the research, extension, client and input sub-systems.
4) To suggest a model of linkages between and among the research, extension, client and input sub-systems in the TOT in agriculture.

With a view to studying the structural linkages between and among the four sub-systems involved in the TOT process, a detailed analysis of the formal mechanisms of linkage in the KAU, Kerala State Department of Agriculture, Farmers' organisations and input agencies was made using review of literature and personal discussions with the representatives of the four sub-systems.

To study the functional linkage, two agro-climatic regions of the State viz. central region and region of problem areas which formed the service areas of the RARS Pattambi, RRS Vyttila, Moncompu and

Kayamkulam were selected. ARS Mannuthy was also included since it is functioning as a sub-centre of the central region and region of problem areas.

Since the study aimed at systems analysis of existing linkages for TOT, respondents were drawn from all the four sub-systems viz. research, extension, client and input sub-systems.

The RSS consisted of all the scientists of the KAU who had been associated with evolving and releasing improved rice varieties from the Rice Research Stations at Pattambi, Vyttila, Moncompu and Kayamkulam. Out of the 55 scientists in position, only 52 of them were available at the time of data collection and they formed the respondents of the RSS.

The ESS consisted of the officers of the Department of Agriculture selected from the sub-division to village level, working in the eight sub-divisions viz. Alleppey, Chengannur, Mavelikkara, Alwaye, Shoranur, Mannarghat, Chittur and Alathur. While in the cases of SDAOs and SMSs, complete enumeration was done, probability proportionate to size random sampling procedure was adopted in respect of the AOs and ADs. Thus there were seven SDAOs, 28 SMSs, 56 AOs and 43 ADs making a total of 134 respondents from the ESS.

Contact farmers in the eight sub-divisions, who had tried at least once and continued to grow the improved rice varieties released by the KAU formed the CSS. Two contact farmers were selected at random from the 55 AE units of the eight sub-divisions. Thus there were 110 respondents from the CSS.

Dealers of fertilizers, plant protection chemicals and equipment and credit institutions of Alleppey, Ernakulam, Trichur and Palghat districts constituted the ISS. The total number of respondents from the ISS was 32. Thus the final sample included 52 respondents from the RSS, 134 from the ESS, 110 from the CSS and 32 from the ISS making an aggregate of 328 respondents for the study.

Since the major statistical method applied in the research was factor analysis, there was no categorisation of variables as dependent and independent. But for factor analysis purpose, the 12 possible combinations with regard to the extent of linkages between and among the four sub-systems were considered as variables. Under each variable, all the activities that could possibly establish a linkage between two sub-systems were selected based on the relevancy rating by the judges. Similarly, information on constraints in evolving, spreading and adopting improved rice varieties released by the KAU was also gathered from the sub-systems.

Data from the RSS, ESS and ISS were collected using separate pre-tested questionnaires. Data from the CSS were collected using a pre-tested structured interview schedule. Data were collected during September-December 1987.

The statistical tools used were mainly principal factor analysis and the Friedman two-way analysis of variance by ranks test.

The salient findings of the study were as follows:

1. The overall analysis of existing structural linkages in the TOT process pointed out to the existence of appreciable linkage forums to facilitate effective TOT of improved rice varieties released by the KAU, particularly with reference to the RSS and the ESS. The study also revealed the lack of organised linkage forums for the ISS and the CSS.

   It was found that there was no representation for the ISS in the important linkage forums of the RSS like the Research Council, the Research Advisory Committee and the Extension Advisory Committee. Similarly, representation for the CSS is not provided in the Package of Practices Workshop, the Research Advisory Committee and the Zonal workshops of NARP and T&V system.

   Representation of the ESS in the above linkage forums is found to be restricted to the officials of the cadre of SMS and above.

   Representation for the CSS is found lacking in the important linkage forums of the ESS like the SLTC and the Fertilizer standing committee.
2. The extent of linkage of the RSS with the ESS was on par with that of the CSS. But the extent of linkage of the RSS with the ISS was significantly poor when compared to that of the ESS and the CSS.
3. The extent of linkage of the ESS with the CSS was on par with that of the ISS. But the extent of linkage of the ESS with the RSS was significantly poor when compared to that of the ESS and the CSS.
4. The extent of linkage of the CSS with the RSS was on par with that of the ISS. But the extent of linkage of the CSS with the ESS was significantly better when compared to that of the RSS and the ISS.

5. There was no significant difference in the linkage activities of the ISS with the RSS, ESS and CSS indicating that the extent of linkage of the ISS with the RSS was on par with that of the ESS and the CSS.
6. With a view to segregate the factors influencing the extent of linkage between and among the four sub-systems, principal-factor analysis was applied and the results are summarised as follows:

   6.1 The factors associated with the linkage of the RSS with the ESS were mainly providing consultancy service and training to the ESS, facilitating village level extension activities, supervising trials and demonstrations, facilitating as well as guiding in field level extension activities, providing information support on improved rice varieties and defending in the package of practices workshops.

   6.2 The factors associated with the linkage of the RSS with the CSS were mainly organising farmers' group discussions and trainings and liaising with farmers' associations, communicating information on rice through print media, organising farmers' days and clarifying the doubts of farmers on rice, and visits to farm trials.

   6.3 The factors associated with the linkage of the RSS with the ISS were mainly enabling the ISS to participate in extension activities of research station, facilitating the extension activities of the ISS, keeping liaison with input agencies and communicating with them.

   6.4 The factors associated with the linkage of the ESS with the RSS were mainly building up knowledge through participation in workshops and joint field visits, seeking help of researchers in trials and demonstrations, updating knowledge by participating in trainings, workshops etc. and facilitating research by suggesting new area of research.

   6.5 The factors associated with the linkage of the ESS with the CSS were mainly liaising with farmers associations and motivating them, supervising the farm trials, advising and enabling the farmers on rice growing, interacting with farmers in seminars, discussions, etc., and communicating through mass media.

   6.6 The factors associated with the linkage of the ESS with the ISS were mainly by way of interacting with them in workshops, seminars, etc. exercising regulatory powers, keeping liaison with the ISS and organising training for them, communicating through publications of the ISS and facilitating farmers by making available credit in time.

6.7 The factors associated with the linkage of the CSS with the RSS were mainly seeking specific recommendations/ solutions to problems on rice from the researchers, seeking incentives of inputs and seeking information on rice through mass media and by way of visits to trials in other farmers' fields.

6.8 The factors associated with the linkage of the CSS with the ESS were mainly seeking knowledge and solutions for problems on rice, seeking information on rice through mass media; interacting with the ESS in various discussion forums, and co-operating with the ESS in farm trials.

6.9 The factors associated with the linkage of the CSS with the ISS were mainly by seeking knowledge by participating in group discussions, seminars, etc. and co-operating with the ISS in the conduct of these forums; seeking help from them in preparing production plans and communicating with them.

6.10 The factors associated with the linkage of the ISS with the RSS were mainly facilitating some of the research and extension activities of researchers, getting validated the quality of their products through researchers, interacting with researchers in workshops and seeking the help of the researchers in laying out demonstrations.

6.11 The factors associated with the linkage of the ISS with the ESS were mainly co-operating with the ESS in their extension activities, seeking help of the ESS in tackling field problems and interacting with the ESS by participating in campaigns, seminars etc. organised by the ESS.

6.12 The factors associated with the linkage of the ISS with the CSS were mainly product promoting activities like campaigns, exhibitions etc., extending credit facilities to farmers, giving technical information to farmers on inputs and receiving the complaints of farmers and replying them properly.

7. The perception about the importance of linkage activities of the RSS with the ESS was on par with that of the CSS. But the

perception about the importance of linkage activities of the RSS with the ISS was significantly poor when compared to that of the ESS and the CSS.

8. The perception about the importance of linkage activities of the ESS with the CSS was on par with that of the ISS. But the perception about the importance of linkage activities of the ESS with the RSS was significantly poor when compared to that of the CSS and the ISS.
9. The perception about the importance of linkage activities of the CSS with the RSS was on par with that of the ISS. But the perception about the importance of linkage activities of the CSS with the ESS was significantly better when compared to that of the RSS and the ISS.
10. The perception about the importance of linkage activities of the ISS with the RSS was almost on par with that of the CSS. But the perception about the importance of linkage activities with the ESS was significantly better when compared to that of the RSS and the CSS.
11. Identification of constraints in evolving and spreading improved rice varieties indicated lack of infrastructural facilities for research followed by insufficient staff strength and lack of facilities to multiply seeds as the first three important constraints in the descending order as perceived by the RSS.
12 The ESS, CSS and ISS perceived the susceptibility of improved rice varieties to pests and diseases followed by the need for heavy dose of fertilizers and pesticides and poor cooking quality as the first three constraints in the descending order, in spreading and adopting the improved rice varieties released by the KAU.

Implications of the Study

1. The index developed in the study to quantify the perception of the respondents regarding the importance of the linkage activities, holds out definite scope for further improvement and for use in future research studies of similar nature.
2. The results of the study pertaining to the structural linkage indicate that there was inadequate linkage forums in the case of the ISS and the CSS. Even in the case of the RSS and the ESS where adequate linkage forums exist, there was lopsided representation provided for important functionaries of the ISS and

the clientele. Yet another notable feature was that in the structural linkage forums of the RSS and the ESS, representation to personnel of the lower rung of the hierarchy is totally lacking. Besides, the periodicity and regularity of meeting of these structural forums leave much to be desired. These call for a thorough analysis to provide cross-sectional representation in the structural linkage forums like package of practices workshops, zonal workshops, EAC, SLTC etc. for the sub-systems studied.

3. In the case of the CSS, because of its unorganised nature and the over-politicisation of farmers associations, genuine structures for facilitating linkage with RSS, ESS and ISS seem only a distant possibility.
4. To ensure effective linkages between and among the RSS, ESS and CSS, validation programmes such as the farm trials should be effectively and jointly carried out on farmers' fields with the active participation of these sub-systems. As found out in this study, supervision of these farm trials by the RSS and ESS was inadequate due to lack of facilities for mobility and other resources. It is therefore, suggested that adequate facilities should be provided for the RSS and ESS for undertaking field visits.
5. One of the determining factors of linkage between the RSS and the ESS is the participation of the ESS in trainings conducted by the RSS. The recent move of the Department of Agriculture to sponsor only the senior level extension personnel for training organised by the RSS has deprived the lower level extension functionaries of opportunities to liaise with the RSS effectively. This has to be rectified by deputing all the categories of extension personnel of the Department of Agriculture for training organised by the KAU.
6. The constraints identified indicated that rice research and extension in the State are confronted with multiplicity of problems including lack of research infrastructure and facilities for better farmer contact programmes. Facilities for multiplication of seeds are lacking which may jeopardise the entire TOT process. Therefore, urgent measures may be initiated to strengthen facilities for research, seed multiplication, and for better farmer contact programmes on the part of the RSS.
7. While many commercial crops have specific development boards in the State/National level, rice is one crop which does not have any. A state level coordinating agency in the form of a 'Rice

Development Board' will be a welcome step in this direction.

8. The potential of the mass media like newspaper, radio and television to reach the farming community seems to be the lost sight of by the RSS, ESS and ISS. This could be due to lack of exposure of these personnel to the effective use of mass media for TOT. Periodic workshops and trainings may be organised to surmount this handicap.
9. The suggested research, extension, client and input sub-systems linkage model emphasising the need for strengthening the identified weak structural and functional linkage activities between and among the four sub-systems could be used with suitable modifications.

BIBLIOGRAPHY

Allen, Richard K. (1977). *Organizational management through communication*. Harper and Row, New York.

Amalraj, V.A. and Prasad, M.S. (1984). Role of fertilizer dealers in the transfer of new farm technology. *Indian J. Extn. Edn., 20* (3 & 4) : 69-70.

Ambastha, C.K. (1980). Communication pattern of farm scientists. *Indian J. Extn. Edn., 16* (1&2) : 34-38.

——— (1986). *Communication patterns in innovation development, extension and client systems*. B.R. Publishing Corporation, Delhi.

——— and Singh, K.N. (1976). Farm scientists communicate to farmers. *Indian J. Extn. Edn., 12* (3&4) : 1-5.

——— (1977). Communication pattern of farm scientists: A systems analysis. *Indian J. Extn. Edn., 13* (1&2) : 22-30.

Anantrao, N.K. (1988). New goals for farm research. In *Survey of Indian Agriculture, The Hindu*, Madras.

Anonymous (1969). *Evaluation report of the HYVP: Report for rabi 1968-69—Paddy and Jowar*. Programme Evaluation Organisation, Planning Commission, New Delhi.

——— (1976). *Report of the National Commission on Agriculture*. Govt. of India, Ministry of Agriculture & Irrigation, New Delhi, Part XI.

——— (1977). *Constraints to high yields on Asian rice farms: An Interim report*. International Rice Research Institute, Philippines.

——— (1978). *Report of the Review Committee on Agricultural Universities*, ICAR, New Delhi.

——— (1982). Recommendations of the national seminar on challenges of extension in eighties. *Indian. J. Extn. Edn., 18* (1&2) : 26-33.

——— (1984). National seminar on professional agricultural extension system. In *Agricultural Situation in India, 38* (12) : 834.

Asawa, B.M. (1981). Factor analysis in chickpea. *Indian J. Agric. Sci., 51* (37) : 156-157.

Awa, Njoku and Crowder, L. Van (1978). How extension stacks up? *J. Extn., 17:* 19-25.

Axinn, George H. and Thorat, Sudhakar (1972). *Modernising world agriculture — A comparative study of agricultural extension systems*. Oxford & IBH Publishing Co., New Delhi.

Baird, John E. Jr. (1977). *The dynamics of organizational communication*. Harper and Row, New York.

Beal, G.H. and Rogers, E.M. (1959). The scientist as a referent in the communication of new technology. *Publ. opin. Qrtly., 22:* 559.

Benor, Daniel, and Baxter, Michael (1984). *Training and Visit Extension*. World Bank, Washington, D.C.

Benor, Daniel and Harrison, James Q. (1977). *Agricultural Extension: The Training and Visit System*. World Bank, Washington, D.C.

Bhatnagar, O.P. and Desai, G.R. (1987). Management of agricultural extension: Concepts and constraints. *J. Rur. Devt., 6* (1): 1-66.

Bhatnagar, O.P., Desai, G.R. and Reddy, M.R. (1986). Management of agricultural extension under T & V system in India: An over view. *J. Rur. Devt., 5* (5): 561-573.

Blalock, Hubert M. Jr. (1960). *Social Statistics*. McGraw-Hill Book Company, New York.

Buckley, Walter (1964). *Sociology and modern systems theory*. Prentice-Hall Inc., London.

Chand, Jain Nemi (1971). Communication pattern and effectiveness of professionals performing linking roles in research dissemination organisation. Ph.D. Thesis, Michigan State University, USA.

Chatterjee, Biswa B. (1974). Learning and decisions: Two aspects of transfer technology. In Daulat Singh (Ed.), *Transfer Technology: Systems and Constraints*, U.P. Society of Extension Education and Rural Development, Kanpur.

Coombs, Philip H. and Ahamed, Manzoor (1974), *Attacking rural poverty*. Johns Hopkins University Press, London.

Desai, G.R. (1987). Linking research and extension for agricultural development — The case of Karnataka. *J. Rur. Devt., 6* (3): 272-279.

Dubey, V.K. (1974). Systemic constraints involved in the process of technology dissemination. In Daulat Singh (Ed.), *Transfer Technology: Systems and Constraints*, U.P. Society of Extension Education and Rural Development, Kanpur.

Dudhani, C.M. (1980). A study on the linking roles of extension supervisors, SMS and research workers. *Ph.D. Thesis*, University of Agricultural Sciences, Bangalore.

Dwarakinath, R. and Channegowda, M.P. (1974). Constraints to the transfer of technology. In Daulat Singh (Ed.), *Transfer Technology: Systems and Constraints*, U.P. Society of Extension Education and Rural Development, Kanpur.

Edward, D.E. (1957). *College for our land and time*. Harper and Row, New York.

Eklund, P. (1985). Technology development and adoption rate: systems approach for agricultural research and extension. *Agricultural Research Management Abstract, 1* (1): 47.

Ensminger, Douglas (1986). For a first step technology. *Ceres, 19* (3): 6.

*Friedman, M. (1937). The use of ranks to avoid the assumption of normality implicit in the analysis of variance, *J. Amer. Statist. Assn., 32*: 675-701.

Fruchter, Benjamin (1954). *Introduction to factor analysis*. Affiliated East West Press Pvt. Ltd., New Delhi.

Garg, J.S. and Omprakash (1974). Techno-economic barriers of crop production technology of HYV. In Daulat Singh, (Ed.) *Transfer Technology: Systems and Constraints*. U.P. Society of Extension Education and Rural Development, Kanpur.

Gee, Sherman (1981). *Technology transfer, innovation and international competitiveness*. John Wiley & Sons, New York.

Geer, Van de (1971). *Introduction to multivariate analysis for the social sciences*. W.H. Freeman and Co., San Francisco.

*Hall, A.D. and Fagen, R.E. (1956). Definition of system. *General Systems Year Book, 1*: 18-28.

Harman, Harry H. (1960). *Modern factor analysis*. The University of Chicago Press, London.

Havelock, Ronald G., Guskin, A., Frohman, M., Havelock, M., Hill, M. and Huber, J. (1969). *Planning for innovation*. Institute of Social Research, Michigan State University, USA.

Jaiswal, N.K. and Arya, H.P.S. (1981). *Management of transfer of farm technology*. National Institute of Rural Development, Hyderabad.

Jaiswal, N.K. and Das, P.K. (1981). Transfer of technology in rice farming, *Rur. Devt. Digest, 4* (4): 320-353.

Jaiswal, N.K., Roy, N.K. and Singh, B.S.P. (1970). Effects of size of holdings and irrigational facilities on adoption of HYV. *Indian J. Extn. Edn., 6* (1 & 2): 8-15.

Johnson, A., Richard, E.F. Kast E. Fremont, and Rosenzweing, E. James. (1967). *The theory and management of systems*. McGraw-Hill Book Company, New York.

Joon, B.S., Jagadish Singh, and Rana, O.P. (1970). Response of the farmers towards the HYV. *Indian J. Extn. Edn., 6* (3 & 4) : 58-62.

Kaiser, H.F. (1958). The Varimax criterion for analytic rotation in factor analysis. *Psychometrika, 23* (2) : 187-200.

Kamath, K.G. (1974). *Agricultural research and communication in India*. National symposium on agricultural research and development since independence. ICAR, New Delhi: 613-619.

Kaur, Dil Bagh (1985). Problems of transferring technology to rural areas. *Yojana, 29* (22) : 19.

Kelsey, Lincoln David and Hearne, Cannon Chiles (1963). *Cooperative extension work*. Comstock Publishing Associates, New York.

Kerlinger, Fred N. (1978). *Foundations of behavioural research*. Surjeet Publications, New Delhi.

Keuls, M. (1952). The use of the 'Studentized range' in connection with analysis of variance. *Euphytica, 1*: 112-122.

Khan, A.W. (1980). Communicating information to rural people. In *Rural Development and Communication policies*, Institute of Mass Communication, New Delhi.

Kishore, Devesh (1986). An alternate strategy for the transfer of technology with special reference to India. *Agricultural Administration, 21*: 197-204.

Koontz, Harold, Cyril O'Donnell, and Heinz Weigrich (1980). *Management*, McGraw-Hill Book Company, New York.

Leagans, Paul J. (1985). Adoption of technology by small farmers: A Model for strategy builders. *Journal of Extension Systems, 1* (1) : 1-22.

Lowdermilk, Max K. (1985). A system for improving the quality of extension. *Journal of Extension Systems, 2* (1) : 45-53.

Madeuf, B. (1985). International technology transfers and international technology payments: Definitions, measurement and firms behaviour. *Agricultural Research Management Abstract, 1* (1) : 47.

Malekmohammadi, Iraj (1987). Professional competencies of extension personnel in relation to innovation diffusion process in Iran. *Indian J. Extn. Edn., 23* (3 & 4): 63-64.

Mann, Michael (Ed.) (1983). *The Macmillan student encyclopaedia of sociology*. Macmillan Press, London.

Matlock, W.G. (1985). Problems and solutions in agricultural technology: Change or transfer. *Agricultural Research Management Abstract, 1* (1) : 48.

Merton, R.K. (1957). *Social theory and social structure.* Free Press, New York.

Mirdha, N.R. (1974). All India symposium on extension role of Agricultural Universities. *Indian J. Extn. Edn., 10* (1 & 2) : 12-15.

Moseman, A.H. (1970). *Building agricultural research system in developing nations.* Agricultural Development Council INC, New York.

Mosher, A.T. (1975). *Serving agriculture as an administrator.* Agricultural Development Council INC, New York.

Newman, D. (1939). The distribution of range in samples from a normal population expressed in terms of an independent estimate of standard deviation. *Biometrika, 31* : 20-30.

Nweke, F.I. and Akorhe, J.A. (1985). Determinants of adoption of new technology among small holders and implications of transfer programmes — A case study of rice production in the Plateau State of Nigeria. *Agricultural Research Management Abstract, 1* (1) : 49.

Okuneye, P.A. (1985). Farmer contact and the demonstration of improved farming practices in Nigeria. *Agricultural Research Management Abstract 1* (4) : 293.

Overall, John E., and Klett, James C. (1972). *Applied multivariate analysis,* McGraw-Hill Book Company, New York.

Pachico, D. and Ashby, J. (1985). Stages in technology diffusion among small farmers – Biological and management screening of new rice variety in Nepal. *Agricultural Research Management Abstract, 1* (1) : 49.

Pal, T.K. (1975). *Change in rice farming in selected areas of Asia (Orissa).* International Rice Research Institute, Philippines.

Pandey, S.N. and Mishra, R.N. (1984). Communication behaviour of subject matter specialists in the T & V extension system. *Interaction, 2* (1 & 2) : 270.

Parthasarathy, G. (1975). *Changes in rice farming in selected areas of Asia (West Godavari Dist.).* International Rice Research Institute, Philippines.

Pickering, Donald C. (1987). An over view of agricultural extension – its linkage with farm research and the farmer — The world Bank experience, *Journal of Extension Systems, 1* (3) : 4-12.

Prasad, Chandrika (1970). Teaching, research and extension in agriculture — A case study. *Indian J. Extn. Edn., 6* (1 & 2); 73-77.

——— (1984). Transfer of agricultural technology. *Maharashtra J. Extn. Edn., 3*: 103-109.

Prasad, Muraleedhara R. (1983). Comparative analysis of achievement motivation of rice farmers in three States in India. Ph.D. Thesis, University of Agricultural Sciences, Bangalore.

Prasad, V.L. and Reddy, T.J. (1986). Farming system research for livestock development. *J. Rur. Devt., 5* (6): 645-653.

Quade, E.S. and Boucher, W.I. (1968). ***Systems analysis and policy planning: Application in defence.*** American Elsevier Publishing Company, New York.

Rajasekhar, David O. (1984). Yield gap analysis — A study of selected crops in Madurai district. M.Sc. (Ag.) Thesis, Tamil Nadu Agricultural University, Coimbatore.

Rao, Radha C. (1972). Communication linkages in transfer of agricultural technology. *Econ. Pol. Wkly., 7* (53) : 157-170.

Ray, G.L. (1976). Reasons for low spread of HYV of paddy during pre-kharif and kharif season in West Bengal. *Indian J. Extn. Edn., 12* (1 & 2) : 59-64.

Reddy, Adivi A. (1981). Essential linkages for effective transfer of agricultural technology: A systems approach. In *Management of Transfer of Farm Technology*. National Institute of Rural Development, Hyderabad.

Reddy, Pandarinath (1986). Extension systems interaction with research and client systems — An inter-system analysis. *Journal of Extension Systems, 2* (2) : 36-42.

Rogers, Everette M. (1983). *Diffusion of innovations*. The Free Press, New York.

——— and Shoemaker, F.F. (1971). *Communications of innovation: A cross-cultural approach*. The Free Press, New York.

——— and Yost, M.D. (1960). Communication behaviour of county extension agents, Wooster. *Ohio Agricultural Experimental Station Research Bulletin:* 850.

Sabarathnam, V.E. (1985). Heterogeneity of dissemination linkages among ARS scientists. *Paper presented at the National Seminar on management of agricultural and rural development programmes.* College of Agriculture, Nagpur.

Sadikin, S.W. (1985). Overcoming technology gaps. *Agricultural Research Management Abstract, 1* (1): 51.

Samantha, R.K. and Kishore, Devesh (1984). New dimensions of transfer of agricultural technology. In *Changing perspectives in extension* (Mimeo). National Institute of Rural Development, Hyderabad.

Sampath, K., Pannirselvam, A. and Santhanam S. (1981). *Introduction to educational technology*. Sterling Publishers Pvt. Ltd., New Delhi.

Schumacher, E. (1973). *Small is beautiful*. Blond and Briggs, London.

Sen, D. (1984). Changing perspectives in extension: Some relevant issues. In *Changing perspectives in Extension* (Mimeo). National Institute of Rural Development, Hyderabad.

Sethy, B., Singh, P.P. and Bahal, R. (1984). Some interpersonal characteristics in adoption of farm technology. *Indian J. Extn. Edn., 20* (1 &2) : 29-37.

Sharma, J.C. (1982). Linkages among research, extension, farmers and input and credit organisations at various levels. *Indian J. Extn. Edn., 18* (1 & 2) : 16-24.

Sharma, J.S. (1975). *Changes in rice farming in selected areas of Asia (UP)*. International Rice Research Institute, Philippines.

Sharma, S.K. and Nair, G.T. (1974). A Multivariate study of HYV of paddy. *Indian J. Extn. Edn., 10* (1 & 2) : 30-35.

Singh, Daulat (Ed.) (1974). *Transfer technology, systems and constraints*. U.P. Society of Extension Education and Rural Development, Kanpur.

Singh, Indu B. (1980). Current perspectives in mass communication theory and research. *Indian J. Extn. Edn. 16* (3 & 4) : 53-66.

Singh, K.N. (1984). Developing linkages between research, extension and farmers. In *Changing Perspectives in Extension* (Mimeo). National Institute of Rural Development, Hyderabad.

Singh, K.N. and Prasad, N. (1970). A Study on coordination between cooperation and agricultural departments. *Indian J. EXtn., 6* (1 & 2): 1-7.

Singh, Krishna B, Lin Corbin and Marsha Wadkins (1975). Systems theory and techniques of systems analysis. *Indian J. Extn. Edn., 11* (3 & 4); 15-16.

Singh, Labh and Sharma, R.K. (1986). Rice production system — Constraints causing concern to paddy growers. *Journal of Extension Systems, 2* (2) : 32-35.

Singh, S.N. and Babu, Remesh, A (1984). Multimedia approach for accelerating the process of agricultural development. In *Changing Perspectives in Extension* (Mimeo). National Institute of Rural Development, Hyderabad.

Stinchcombe, A.L. (1968). *Constructing social theories:* Har-Court, Bruce and World Inc., New York.

Surendran, G. (1982). Impact of operational research project on agricultural production. M.Sc. (Ag.) Thesis, Deptt. of Agrl. Extension, College of Agriculture, Vellayani, Kerala.

Swaminathan, M.S. (1988). Challenges ahead: Time for new orientation. In *Survey of Indian Agriculture, The Hindu*, Madras.

Thurstone, L.L. (1947). *Multivariate factor analysis.* University of Chicago Press, Chicago.

Tripathy, A., Singh, K.N. and Sankar Sahoo (1982). Constraints in adoption of high yielding rice technology in coastal Orissa. *Indian J. Extn. Edn., 18* (1 & 2): 51-58.

Venkataraman, A. (1988). Gaps in extension service. In *Survey of Indian Agriculture. The Hindu*, Madras.

Verma, O.S. (1974). Problems of transfer of technology and distance as a measure of the system. In Daulat Singh (Ed.), *Transfer Technology : Systems and Constraints*. U.P. Society of Extension Education and Rural Development, Kanpur.

Walton, P.D. (1972). Factor analysis of yield in spring wheat, *Crop Science,* (12) : 731-736.

Zar, Jerrold, H. (1981). *Bio statistical analysls*, John Wiley Sons, New York.

INDEX